5/5/14

PARTNERS, GUESTS, and PARASITES

ALSO BY HILDA SIMON
Feathers, Plain and Fancy
Insect Masquerades

PARTNERS, GUESTS, and PARASITES

Coexistence in Nature

by Hilda Simon

THE VIKING PRESS   NEW YORK

First Edition

Copyright © 1970 by Hilda Simon
All rights reserved
First published in 1970 by The Viking Press, Inc.
625 Madison Avenue, New York, N.Y. 10022
Published simultaneously in Canada by
The Macmillan Company of Canada Limited
Library of Congress catalog card number: 71-106924
574.5 Ecology
591   Zoology
Printed in U.S.A.
Trade 670-54086-2
VLB 670-54087-0
1 2 3 4 5   74 73 72 71 70

Acknowledgments

I gratefully acknowledge the aid extended to me during research for this book by the curators and staff of the Ornithological Department of the American Museum of Natural History; Charles H. Rogers, curator of the Princeton University Museum of Zoology; and Mrs. Dorothea Richter, of the Max Planck Institute at Frankfurt-on-the-Main, Germany.

I also wish to thank Mrs. Beatrice Rosenfeld, science editor of The Viking Press, for her generous help with the manuscript and her invaluable advice.

Contents

Living Together	11
Guests, Passengers, and Parasites	23
Partners for Gain and Convenience	59
Cooperation for Survival	89
Index	123

PARTNERS, GUESTS, and PARASITES

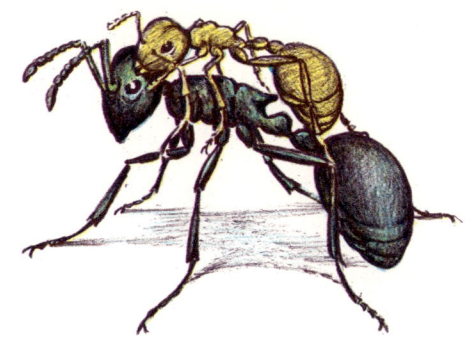

Living Together

At the heart of Charles Darwin's theory of evolution is the premise that all evolutionary development is based upon the constant and unrelenting struggle for survival. Many of Darwin's followers, duly impressed by the fact that preservation of the species seems to be one of the main goals in nature, overemphasized the aggressive aspects of the struggle. Animal life was pictured as "red in tooth and claw," although Darwin never implied that the struggle for survival necessarily has to involve bloodshed and violence. Despite new insights into the important role of cooperation in evolution, the one-sided concept of nature as a big bloody battlefield tends to persist. Many books still stress the "food chain," in which the weaker and smaller animals are eaten by the stronger and larger ones. At the same time, associations between species that coexist peacefully side by side are often largely neglected. Together with another one-sided concept that all anatomical features produced by evolution are adaptive, having so-called survival value, this has created a curiously simplified approach to the endlessly varied forms of animal life.

The food-chain relationship is a vital part of nature's household, supplying food to a host of animals while acting as a check on the population of others. However, a great many animals do *not* kill and eat others. An equally impressive number do *not* live in constant fear of being killed, and this applies not only to the large and powerful creatures such as the gorilla, the elephant, the buffalo, and the big cats that are normally immune to attack by others. Many small, weak animals have carved out their own modest, noncompetitive niches in which they live in comparative security. Moreover, many little-understood relationships involving mutual assistance between organisms in both the plant and animal worlds seem to be necessary for maintaining the delicate, dynamic balance of nature, and apparently these relationships cannot be greatly disturbed without setting off a chain reaction of detrimental effects.

Appreciation of the need to maintain this balance has become widespread during the last decade. In the past, man was so engrossed in his attempts to conquer and control his environment that he did not consider the long-range effects of his activities. Until quite recently, in fact, the earth's natural resources, including plant and animal life, were taken for granted and assumed to be more or less unlimited. As a result of this attitude, these resources have dwindled at a frightening rate—especially in the past fifty years—and it has become obvious that drastic steps must be taken to remedy the situation before the point of no return has been reached. Hence ecology, the relatively new branch of biology which embraces the entire field of interrelationships between living organisms and their environments, is today receiving increased attention.

Even a superficial study of ecology quickly establishes the fact that a fantastic and colorful variety of often complex associations is found in nature. Many of these associations are peaceful, ranging from indifferent tolerance to active cooperation between two entirely different types of organisms. Such a relationship is known as *symbiosis*, a term derived from Greek, meaning "living together." However, the word is never applied to associations among animals of the same species, such as a herd of elephants, a flock of geese, a school of fish, or a colony of ants. This term is used exclusively for relationships between two different kinds of animals or organisms—a bird and a mammal, for instance, or two species of fish, or a plant and an insect.

Although the term "symbiosis" was introduced less than a hundred years ago by a French biologist to describe the association between an alga and fungus that together form a lichen, interest in such relationships has long existed. Since ancient times, naturalists have commented upon some of the unusual "friendships" between two different creatures. Among the first of the more modern observers who studied this phenomenon was a German schoolteacher, Christian Konrad Sprengel. In 1793 he wrote a book with the awkward title *The Discovered Secret of Nature in the Structure and Fertilization of Flowers,* in which he discussed the symbiotic relationships between certain insects and plants. At the time, however, such treatises, which clashed with established religious concepts, were discouraged by school and clerical authorities. Sprengel was soon forced into retirement, and thereafter had to eke out a meager living by private tutoring.

Decades later Darwin read Sprengel's book with great interest, and the theories of the German naturalist influenced his own ideas on evolution. Fascinated by the evidence of symbiosis in general, and by the results of his experiments with ants and aphids in particular, Darwin concluded that "although there is no evidence that any animal performs an act for the exclusive good of another . . . yet each tries to take advantage of the instincts of others." This, in a nutshell, is an excellent description of the typical symbiotic relationship.

In its broadest sense, symbiosis comprises all types of close associations between two different organisms, including those of a parasitic nature, which are known as *antipathetic* or *antagonistic* symbioses. In its narrowest sense, the word is used to describe only those relationships that are mutually beneficial to both "partners." Between these two extremes lies a whole range of symbiotic associations that are fascinating in their variety.

Parasitism is a much more widespread phenomenon in nature than is commonly known. In fact, almost half of all known animal species are parasitic. By definition a parasite is an organism—plant or animal—that gains food, and often shelter and other advantages, at the expense of its host without the use of violence and without giving anything in return. While the host could live very well—indeed, much better—without the parasite, the latter usually could not live without the host.

To us the idea of parasitism is repulsive. However, under no circumstances can our moral standards be applied to the behavior of other animal groups. While a biologist may apply the term "degenerate" to a parasite, he is using the word not in a moral sense, but as a biological description of an organism that has returned to a less highly organized condition and has lost the independence—and among animals, the freedom of action—which is the hallmark of advanced members of all groups.

In some cases a single parasite may cause considerable injury to, and eventually kill, the host. The "efficient" parasite, however, does not seriously harm its

From left to right, the sheep tick (a species of fly), a flea, and a tick are examples of ectoparasites.

host through its individual activities. Yet many parasites victimize a single host in such large numbers that their collective activities may cause the host's illness or even death. These are what we might term "incompetent" parasites, for it is in their own best interests to keep the host alive, if not well.

The secondary effects of parasitism pose the most serious threat to many host animals, including man. Lice and fleas, for example, may do comparatively little harm by their own parasitic activities. Yet they can become deadly as carriers of dread diseases, such as bubonic plague, the Black Death of the Middle Ages.

Biologists distinguish between different types of parasites. Lice, fleas, and others with biting or sucking mouthparts are called *ectoparasites*. They live on the exterior of the host animal, as distinguished from *endoparasites*, such as hookworms and tapeworms, which live inside their hosts. Endoparasites have evolved special anatomical features which permit them to adapt to their peculiar way of life, and usually lack other organs that enabled their ancestors to lead an independent existence. They are the most dependent of all parasites, and would quickly die out if host animals were not available.

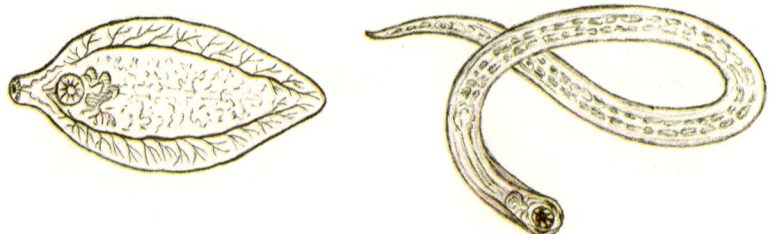

The liver fluke (left) and the hookworm are endoparasites.

The eggs of this ichneumon wasp are deposited in certain caterpillars and develop into larvae that feed upon, and eventually kill, the host insect.

Many insects commonly grouped with parasites are in reality predators, and would be more accurately called *parasitoid*, or parasite-like. In most cases the adults of such parasitoid insects, such as certain wasps, deposit their eggs either on or inside the bodies of other insects that then serve as food for the developing young. In contrast to most true parasites, the direct result of the parasitoid insect's activities is the death of the host animal.

Quite a different type of parasitic creatures is grouped under the so-called social parasites. These are generally found in insect colonies, such as bee hives and ant or termite nests, which are rich and tempting sources of food and shelter if the defenses of the legitimate owners can be successfully evaded. As we shall see later, scores of parasitic insects have managed to do just that, despite the fact that social insects guard their communities jealously against intruders. For instance, a parasitic ant species will manage to take over another ant community and replace the "legitimate" queen and workers with its own. Such examples prove that coexistence among species is not always feasible, and that the normal fierce hostility exhibited by most social insects toward invaders is a necessary self defense.

The specter of parasitism stands behind all symbiotic relationships in which the advantage to one partner might be extended to a one-sided dependency that could be harmful to the other partner. Biologists believe, on the strength of the evidence provided by semi-parasitic associations, that all parasitism evolved from the deterioration of "natural partnerships." They also think they can detect in a number of today's genuine, mutually beneficial relationships certain signs indicating they could possibly turn into parasitism some time in the distant future.

Despite such inherent dangers, the majority of symbiotic alliances are more or less advantageous to both sides, and often indirectly benefit others as well. The ecological importance of certain natural partnerships is well known, though the significance of others has not yet been fully appraised.

Because of the limitations of space, this book will explore only a few examples of various types of symbioses mainly found among the higher forms of life. Of the mutually beneficial relationships between various microorganisms on one hand, and higher plants and animals on the other, suffice it to note that the popular concept associating all bacteria with disease, decay, and death is quite erroneous. While we are more or less aware that the normal intestinal flora and fauna of animals, including man, not only does no harm, but is indeed beneficial, few people know that many plants

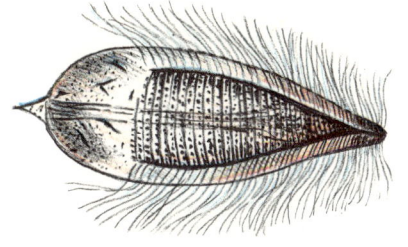 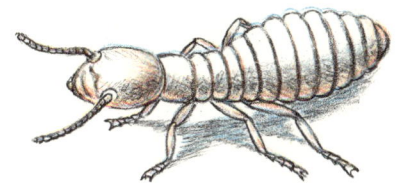

A complex one-celled animal earns its "board and lodging" in the intestines of wood-eating termites (such as the one on the right) by converting indigestible cellulose into digestible matter.

and animals depend upon microorganisms for survival. There are animals that could not exist without the active aid of bacteria they harbor in their bodies. Termites, for instance, as well as other wood-eating insects, have bacteria living in their intestinal tracts. These bacteria secrete enzymes which break down cellulose and thus make it possible for such insects to digest wood, their staple food. This is a true symbiotic relationship, even though we may find it a bit difficult to think of bacteria in terms of "partners." And it is also an example of a symbiotic association which indirectly benefits the environment. While we usually think of termites only as pests, they are a necessary part of the community in many areas uninhabited by man. By eating dead wood, they permit the chemical compounds in the wood to return quickly to the soil as usable products, and also help prevent young plant growth from being choked to death under steadily accumulating piles of dead trees and underbrush.

Not only insects, but large animals such as cattle and other ruminants also depend upon their minute bacterial "partners" to break down cellulose and convert it into a digestible form. The role of man's intestinal microorganisms, while still insufficiently understood, is much more important than had been previously assumed, as it is now known that some of these bacteria manufacture vitamins and other nutrients essential to good health.

As the study of ecology progresses, it provides increasing evidence for a variety of symbiotic associations that form a counterbalance to the aggressive and violent side of nature which has for so long been emphasized. On this other, gentler side of nature we find peaceful coexistence, and even active cooperation, between different species. Amazingly enough, such associations are sometimes found to exist between large, normally aggressive predators and weak, defenseless creatures.

As we examine these interesting natural partnerships in detail, it is clearly demonstrated that the struggle for survival in nature is not always as grim and violent as some would lead us to believe; it is not all just aggressiveness, hunting and killing, stalking and sudden death; not all weaker animals must "naturally" become the victims of the stronger predator. It is a welcome relief to find that tiny, defenseless creatures will venture, unafraid and quite deliberately, into what appear to be the very jaws of death—and emerge unscathed, because the jaws, though formidable and lined with razor-sharp teeth, happen to belong to the tiny creature's huge "partner." And the little animal knows quite well, through whatever instinct supplies this knowledge, that it is quite safe for him to take what looks like a fatal risk. Why and how he knows that, why and how his partner knows when not to clamp his jaws shut and swallow, are just two of the countless riddles and wonders presented by the colorful world of natural partnerships which will be explored in the pages that follow.

Guests, Passengers, and Parasites

One of the most common types of symbiosis is the relationship from which only one partner benefits. Typically the other partner, while not gaining anything from the association, does not stand to lose anything either, and usually adopts an attitude of tolerant indifference toward his companion. There are, however, many borderline cases in which the beneficiary displays parasitic tendencies and a type of behavior that can be classified as social parasitism.

One-sided associations involve animals of almost every description, from mammals to the most primitive creatures, and display a wealth of diversity and interesting peculiarities. Some ties are relatively loose and temporary, others, intimate and permanent. Biologists distinguish between such subdivisions as *commensalism*, *synoecy*, and *phoresy*, to name only three of the more commonly encountered types of one-sided symbioses. Commensalism, a word derived from Latin meaning "at the table," designates an association in which the two partners eat together, sometimes even sharing the same food. Synoecy and phoresy are used, respectively, for the relationships in which one partner shares the other's shelter—his burrow or his nest, for example—and those in which a creature obtains transportation by clinging to his usually larger companion.

Eating Together

Among vertebrates—especially those that live on land—commensalism is perhaps the most frequent type of one-sided symbiosis. It is often a rather loose type of relationship. The smaller partner, a self-invited guest, stays near and sometimes on his indifferent host, gaining various advantages that range from an easily accessible food supply (even scraps that fall from the host's table) to safety from attack by enemies. The larger partner hardly ever benefits but is in no way harmed or inconvenienced by the smaller companion's presence; generally he simply disregards him.

One of the best-known examples of commensalism among land vertebrates is the association between the buff-backed heron, commonly called the cattle egret, and such large herbivorous animals as cattle, rhinoceroses, elephants, and buffalo. The cattle egret is a relatively small heron, standing only about eighteen inches high. Its usually all-white plumage changes during the breeding season, when the bird grows buff-colored feathers on its head, breast, and back.

Throughout Africa and parts of Asia, cattle egrets are a familiar sight. Frequently dozens of these birds are found in the company of grazing mammals. They fearlessly walk among their large four-legged companions, and occasionally even steal a ride on the latters' backs, where their slender white shapes perched on the great dark-gray or brown lumbering creatures create an attractive contrast.

Getting a free ride is not, however, the cattle egret's real reason for associating with its mammalian partners. What makes the relationship worthwhile to the birds are the meals they obtain easily as a result of the mammals' motion, which inevitably flushes all types of insects from the grass. In the hunter's language, they act as "beaters" for the egrets. Following in the wake of their companions' ponderous tread, the birds have an easy time snapping up such insects as grasshoppers, beetles, moths, and flies as they are dislodged from their hiding places in the grass. Occasionally, an egret will pick an insect from the back of an animal on which it is hitchhiking, or from one that is resting. Usually, however, the birds feed on the ground, and any benefit to their companions is incidental.

In recent years cattle egrets have established themselves in other parts of the world, including North America. In the United States they are now found along the Atlantic Coast from Florida to New Jersey. Here they have learned that such machines as lawn mowers and rakers are as effective as large animals in flushing insects from the grass. Cattle egrets therefore

can be observed following lawn mowers just as they would follow a cow or a buffalo.

This habit is of course not unique to the cattle egret. Many insect-eating birds—crows, gulls, cow birds, and starlings—learned a long time ago that following a plowing farmer can be most rewarding, that a rich harvest of grubs and other choice morsels is uncovered as the soil is being upturned. For ages peasants in many countries have been used to having these birds follow them as they plow the land, and have appreciated the services rendered by their feathered partners in destroying great numbers of harmful insects.

Such associations are not a necessity for the partner that benefits from them: all these birds could live very well without the help of their large companions, but they find it more convenient to have their prey flushed from the grass and set up for an easy catch. And of course all kinds of insects, including those that as adults or as larvae feed on the dung of grazing animals, are attracted to herds of cattle and other ruminants.

In selecting a partner for the role of the "beater," birds are not choosy and will frequently use any one of several available animals. In South Africa the insectivorous carmine bee eater employs another bird in this capacity. The "working partner" in this case is the Kori bustard, which belongs to a group related to the cranes but completely adapted to life on dry grassy plains. The largest of the bustards stands about forty inches tall. These birds have large eyes and long legs equipped with the three-toed feet typical of many birds that are competent at walking and running. Despite the fact that they can fly quite well, bustards often choose to outrun their enemies. Shy and wary birds, they usually stay out in the open where predators cannot easily approach them without being seen.

Though not brightly colored, some species of bustards are attractively patterned in gray, brown, black, and white. The Kori bustard is mostly mottled-gray and brown above, and light gray beneath, thus blending in well with its environment. All the more startling is the color contrast supplied by its frequent companion. The carmine bee eater is aptly named, for it does eat bees—among other insects—and its plumage is predominantly a bright red, with the exception of the head and rump which gleam in bluish green hues.

Bee eaters belong to a group of rather small birds that are distinguished by the generally bright colors of their plumage, as well as by their habit of including in their insect diet those with poisonous stings that are shunned by most other birds. Bee eaters have long slender bills, and their elongated tail feathers often make up a good part of their total length, which averages just over twelve inches.

The carmine bee eater, one of the most handsome of the group, preys upon all kinds of flying insects, and especially upon the various members of the grasshopper family and their relatives. For this reason the bird finds the bustard a very handy partner. Grasshoppers are abundant on the plains frequented by the bustard, and are flushed from the grass as the big bird walks along with its slow, measured pace. As quickly as the dislodged insects rise into the air, they are snapped up by the bee eater, which rides comfortably along on the back of the bustard, simply waiting for the food to appear. It does not have to fear much competition from the bustard, for although insectivorous when young, these birds are mostly herbivorous as adults. Thus the field is left to the bee eater, which returns to its convenient perch on the bustard's back after each successful foray.

If a bustard is not available, the bee eater will use some other animal as a combination transporter and beater. Sometimes the substitute will be another bird—they have been observed riding on storks, for instance—or a mammal such as a zebra or a donkey.

To understand why many insect-eating birds find it so advantageous to team up with animals whose activities are likely to flush insects from their hiding places, we must keep in mind that such birds hunt mainly by sight. In contrast to many mammals which have poor eyesight, birds have excellent vision. A hawk can spot a mouse moving in the grass from hundreds of feet above, and an insect-eating bird will instantly snap up even a tiny moving creature. Unnecessary movement is therefore exactly what many insects must avoid in order to stay alive. Camouflaged by colors and patterns, they may sit quietly inches away from a hungry bird and remain undetected—and uneaten—as long as they remain motionless. The slightest movement, however, will seal their fate.

The pompadoured hornbill uses monkeys to flush insects from the branches of jungle trees.

That is where the beater enters the picture. If a well-camouflaged insect, disturbed by the approach of a large animal, makes the mistake of taking flight, it will wind up as a snack for the beater's small companion.

One bird that likes to hunt high up in the trees with the aid of a partner is an African member of a peculiar group called hornbills. As indicated by their name, they have oddly shaped, large bills that look as though the birds are horned. Found only in the Old World, these birds are interesting because of their unusual nesting habits. They build nests in hollow trees, and after the female has laid her eggs, the male fills in the opening of the nesting hole with a kind of cement, imprisoning his mate—and later on the young birds too—in the nest for the entire duration of incubation and rearing of the young. He leaves only a small opening through which he feeds the female and his offspring. Only after the latter are fully grown does the female break out of her self-chosen prison. The entire family then resumes a normal life.

The pompadoured hornbill, named for its crest of white feathers, is called the "monkeybird" by the natives of the upper Congo region because of its association with various primates, especially the long-tailed guenon monkeys. These active creatures move in groups from branch to branch in their search for edible fruit. The hornbill follows them, always staying a few feet below and avidly watching for any insects that are disturbed or dislodged by the monkeys' activities. Such night-flying insects as moths, which rest motionless and well-camouflaged by day and are extremely hard to detect as long as they do not move, fall victim to this mode of hunting.

The natives insist that the hornbill "repays" the monkeys by warning them of approaching danger. However, most naturalists feel that these animals really do not need a sentinel because they themselves are very alert and watchful, and that the association is beneficial to the bird only.

So far the instances of commensalism described have mostly been between large herbivorous and comparatively small insectivorous animals, with the smaller partners benefiting from the relationship. Because there is quite obviously no conflict of interest in associations of this type, it could be assumed that such one-sided symbiotic relationships involving food are limited to situations where the larger partner would never be tempted to consider the smaller companion a tasty morsel. However, as we shall soon see, this would be an altogether erroneous assumption. In nature, things are never quite as simple as that.

One species of African hornbill regularly follows the huge columns of driver ants on their periodical forays. These large, aggresive ants are the most frightening and well-organized predators known to man. As they march along in columns numbering in the millions, the soldiers with their huge, sharp jaws systematically overpower and slaughter every living being they find in their path. Even large animals are not safe when attacked by these fierce ants; there have been stories that, on a few occasions, human beings succumbed to the attacks of these insects.

It is not surprising that small wildlife finding itself in the path of a driver-ant column tries to scramble for safety in a headlong flight. Those that can run, run; those that can fly, fly. In their haste to escape, these animals exercise little care, and thus set themselves up as easy prey for the camp-following hornbills. The birds never seem to eat the ants, being content to use them as beaters for flushing other prey.

Nobody can deny that it would be difficult to find, on land or in water, a group of animals that are collectively more voracious and indiscriminately predatory than the sharks. Although some sharks are harmless, quite unaggressive, and even timid, most members of the group are almost constantly on the prowl and are not at all choosy about their food. They will gobble up garbage tossed from a ship as greedily as they will grab any fish or other marine creature they can catch. It is common knowledge that certain species of sharks are justly among the most feared creatures in the ocean. Yet these rapacious monsters are frequently accompanied by not just one or two "partners," but by entire groups of smaller fish which associate fearlessly with the large predators. In doing so, the little fish not only enjoy a considerable measure of safety—for hardly any smaller predator would care to venture too close to a shark!—but can also count on scraps that are left over from their companion's meals. The most amazing fact about these associations is that, on the basis of all available evidence, the sharks never attempt to attack and eat their smaller companions. On the contrary, they appear to deliberately tolerate them for some still unknown reason.

One of the shark's frequent partners is appropriately known as the pilot fish. It is a common sight to find several of these attractive foot-long, blue-gray and black striped fish swimming a little above and ahead of the fearsome torpedo-shaped predator.

The relationship between pilot fish and sharks is so common, and seems so strange, that it was noted as a curiosity hundreds of years ago by sailors as well as

by naturalists. The "friendship" between the two unlikely partners gave rise to a number of stories, most of them fanciful. However, naturalists more than a century ago supplied very accurate descriptions of the association between the two fishes. And while they could see the reasons for the pilot fish staying close to the shark, they wondered—as we still do today—what kept the shark from eating its small companions. Modern biologists have not yet found a satisfactory answer to this question.

Another frequent companion of many sharks is the remora, commonly called the shark sucker. Actually there are several species of remoras, all with more or less the same anatomical peculiarities and the same habits.

The most outstanding anatomical feature of this fish is a suctorial disk at the top of the head. The disk, which is flattened and looks somewhat like a louvered oval-shaped shutter, is actually a greatly modified dorsal fin. It consists of a number of membranes which, when pressed against a reasonably smooth and flat surface, create many small vacuum pockets and thus permit the remora to cling firmly to a large fish. Usually its choice will be a shark; occasionally it attaches itself to some other species of fish, or even to the hull of a moving ship.

A shark with several remoras attached to its sides or belly is a common sight. In this way the shark suckers gain a threefold advantage: transportation through the water without any effort of their own, safety from other predators, and free food in the form of scraps from the shark's meals. Remoras thus reap even more benefits than the pilot fish from their associations with sharks. While it is generally assumed that the shark does not benefit from the relationship, some biologists tend to think that the remoras may at times remove some parasites from the shark's skin.

It would be logical to assume, in view of the remoras' habits, that these fish are poor swimmers, and therefore have developed the suctorial disks as a necessary aid in moving about. However, that is not so, for remoras are very good and fast swimmers. Here, as with

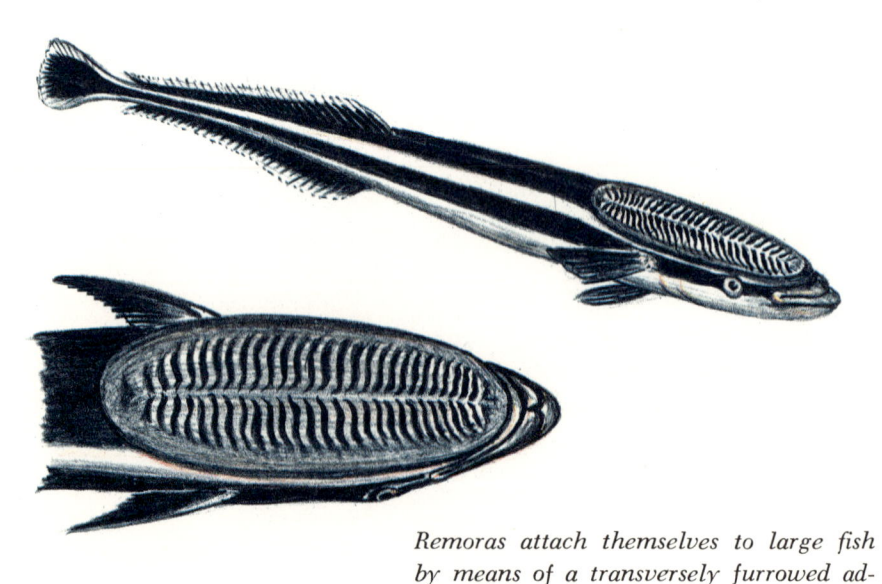

Remoras attach themselves to large fish by means of a transversely furrowed adhesive disk on the head.

the carmine bee eater, the transportation is a convenience and not a necessity. Undoubtedly the safety factor is most important, for remoras attached to a shark or any other large fish are less exposed to danger. Thus, the remoras have found a way of transportation that combines safety with comfort.

Sharing Shelter

Using another creature's abandoned burrow or nest is a common habit among many animals. Moving in with the legitimate owner is another matter. Among land vertebrates this is a comparatively rare form of symbiosis. One of the better known of these associations involves a bird and a reptile. Called the sooty shearwater, the bird, which is related to the petrels and albatrosses, industriously hollows out a comfortable burrow and uses it as a shelter mainly at night. During the day, it is mostly out hunting for food.

The shearwater frequently has an uninvited—and nonpaying—subtenant, a reptile known as a tuatara. This peculiar, slow-moving creature is the only surviving member of a very ancient order. A nocturnal animal, the tuatara usually stays in the burrow during the day and goes out at night to hunt for food, which consists mainly of insects and other small creatures such as worms and centipedes. Because of their respective habits, the two animals are rarely in the burrow at the same time. However, there is no evidence to show that the bird objects to its subtenant even on those occasions when they do meet.

Another bird, which frequently is host to not just one, but two or three subtenants, is the fish hawk, or osprey. This bird of prey builds a huge platform nest of twigs and sticks, in which (or to be accurate, *on* which) it raises its young, usually numbering two. In many cases, however, the young ospreys are not the only bird offspring to be raised in that nest, for very often grackles, night herons, and sparrows will build their own nests right into the sides of the osprey's large structure. The nest then takes on the appearance of an avian apartment building, with the owner occupying the penthouse, while the other floors are taken over— rent free of course—by the various tenants. The landlord ignores them, and they contribute nothing whatsoever to his welfare. On the contrary, they may gain more than ready-made shelters and protection against enemies, for it seems likely that they forage for scraps of food in the osprey's own nest while the adult birds are away hunting.

Other birds will select much stranger landlords. Certain Australian parakeets, for example, build their nests in the large clay mounds erected by the termites of those regions. They do not eat the termites; their food consists mainly of seeds and grain. However, the construction of the nesting hole by necessity destroys part of the termites' carefully laid out runways and chambers, and undoubtedly kills many termites. In spite of this, the termites do not seem to be unduly bothered by the presence of the birds in their midst.

Perhaps even stranger is the case of an Indian wood-

The Indian rufous woodpecker makes its own nest in the nest of the stinging tree ant.

pecker and its hosts, for here two natural antagonists are peacefully united under the same roof at certain times. Like many others of its kind, the woodpecker's diet consists mainly of insects such as ants. Normally it will relish eating the fierce, stinging tree ants of those regions, which build their big, football-shaped nests high up in the branches of large trees. Normally the ants will attack any creature they can overpower. They are especially dangerous to young birds, which they sting to death in their nests. At certain times, however, both the woodpecker and the ants seem to undergo a change of personality and do what ordinarily would *not* come naturally. During this period the woodpecker refrains from eating the ants and the brood of a certain colony, while the ants of that colony refrain from attacking the woodpecker's young, even though both have easy access to each other's offspring.

The phenomenon takes place during the bird's breeding season. At that time a woodpecker will hollow out a nesting hole in the side of one of the large tree-ant nests. In the process, part of the ants' chambers, along with the ants and larvae, are inevitably destroyed. Inexplicably the ants do not retaliate, despite the fact that they are an unusually vicious and aggressive species. The woodpecker proceeds to lay its eggs and raise its young in the ant colony without being harmed by the fierce insect hosts. In exchange the woodpecker parents never touch the ants or any of the brood that belong to the host colony. They will, however, eat tree ants from other communities if they can get hold of any. This is another puzzling instance of symbiosis, and biologists have not yet found a satisfactory explanation for this strange behavior, which appears to be a reversal of both partners' usual habits.

Another extremely strange case of synoecy, although in this instance definitely bordering on parasitism, is the intimate relationship between certain small fish and various ocean creatures, including relatives of the starfish and the pearl oyster.

Fierasfer acus is the scientific name of a fish that is hardly known except to biologists. It has a peculiar development, for it is one of the relatively few fishes that begin life as larvae. We usually think of larvae as being the immature stages of insects, and few people associate the term with vertebrates. However, zoologists use the word for all early stages of those animals which are unlike their parents when young and must undergo a period of bodily changes in order to acquire the adult characteristics.

Eels have this type of development, and so has *Fierasfer acus*, which as a larva lives in the water like other fish. After a while, however, it seeks out a primitive ocean creature popularly known as a sea cucumber, a name that is actually very descriptive for at least certain species of these animals. Called holothurians, sea cucumbers form a class in the larger group of *echinoderms*, meaning literally "prickly skinned." They often have rough or spiny skins and small tubular processes called tube feet, which act like suction cups and which permit the animals to move about.

Sea cucumbers vary in size from just a few inches to almost two feet long. They are usually sausage-shaped with an opening at either end. It would be difficult to tell the front end from the rear if it were not for the small tentacle-like appendages around the "mouth." With these they snuffle around in the mud,

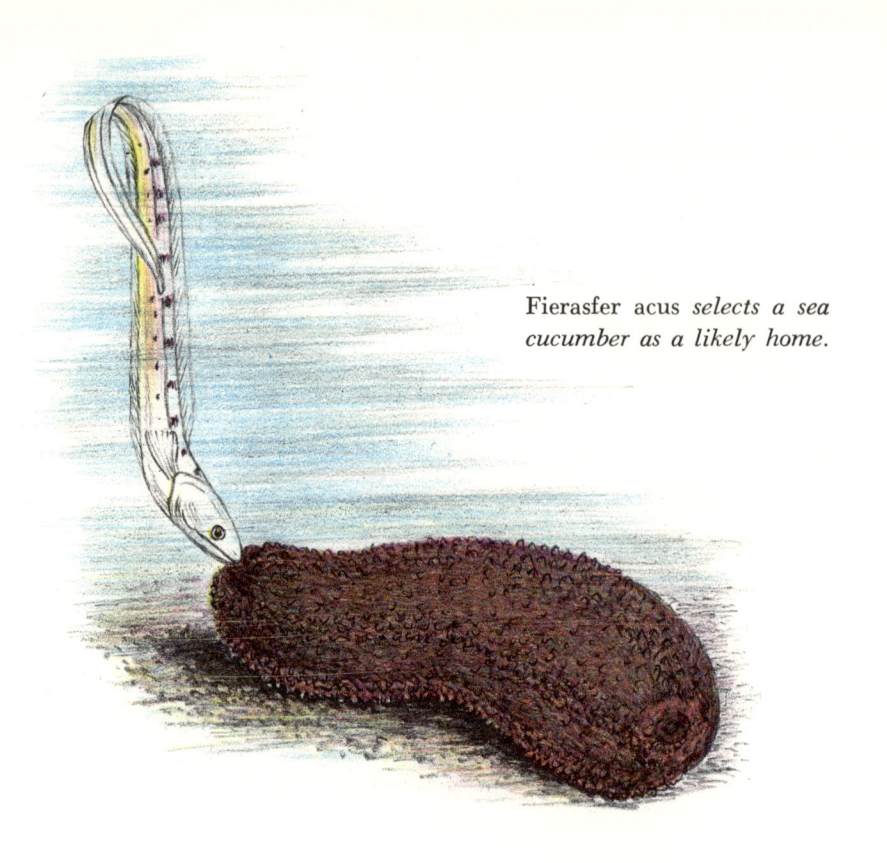

Fierasfer acus selects a sea cucumber as a likely home.

scooping it up and sucking it in. In this way they get their food, which consists of small marine organisms. If the creatures are disturbed, they pull in their tentacles and eject not only the contents of their intestines, but part of the intestinal tract itself. Losing part of their digestive apparatus does not bother them in the least, for like many primitive creatures, they simply grow new parts to replace those that are lost.

This then is the creature that *Fierasfer acus* selects as a host and a home. The fish is wholly dependent upon the sea cucumber, and has special anatomical adaptations that permit it to live inside the holothurian. In its larval stage, after locating a suitable sea cucum-

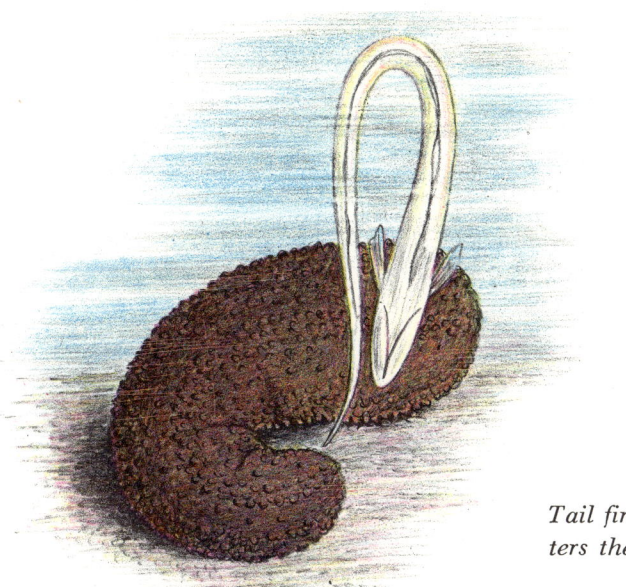

Tail first, Fierasfer *enters the sea cucumber.*

ber, the fish pokes its head inside to see whether it has found the right host, and then turns around and backs in tail first through the rear opening. Being very thin and almost needle-shaped, it fits neatly into the long, narrow, tubelike intestinal tract. With its head usually sticking out and the rest of its body hidden inside the sea cucumber, *Fierasfer acus* spends the remainder of

Once safely inside, Fierasfer *lives inside the body of its host, emerging only occasionally.*

its larval stage well protected against its usual enemies. After the fish is fully grown, it may leave its host from time to time but will always return quickly to the safety of its "living home."

Apparently the sea cucumber does not always feel comfortable with its live-in partner: it seems that at least occasionally it tries to evict the fish when the non-paying tenant does not behave like a polite guest. There is evidence that the fish sometimes pushes through the walls of the intestines and invades the body cavity of its host. Once there, it proceeds to eat some of the latter's organs, which is decidely poor etiquette. However, as was mentioned before, the sea cucumber can easily replace the lost parts and is not greatly harmed by its tenant's uncivil behavior.

Another species of this same group of fishes is called the pearlfish, from its close association with oysters. The little fish has a very convenient arrangement with these shellfish, for it lives inside the mantle of large pearl oysters, where it is safe from many of its enemies and apparently does not disturb the legitimate owner.

There are several other species of fish that live in symbiotic associations with large shellfish. All of them are small, naturally, and none of them seemingly causes any injury to its host. This also holds true for various small species of shrimp, some of which live among the gills in the mantle cavity of certain large mollusks. Because they spend their entire lives hidden from view, they are rarely observed and little known. Some, however, are most attractive. The inch-long

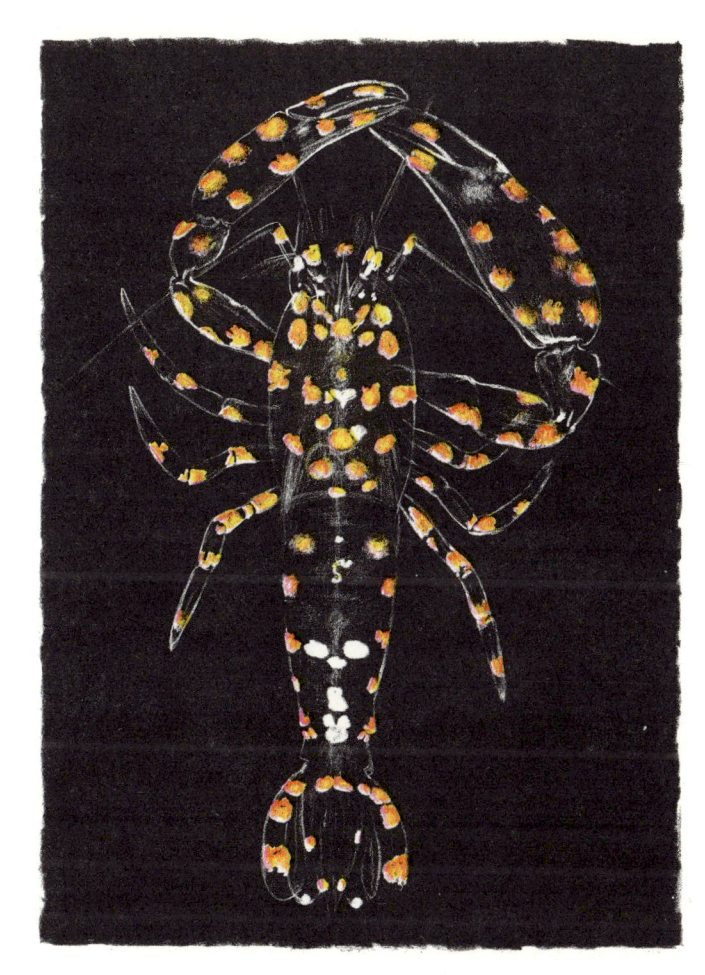

Only about one inch long, this shrimp lives in the mantle cavity of large mollusks, where its transparent, gold-flecked beauty is normally hidden from view.

shrimp *Pontonia*, for example, is practically transparent, but dotted copiously with what look like spots of gleaming gold.

Among the most interesting of small fishes living in symbiotic relationships with other marine animals are the knife-thin urchin fish. Restricted to propelling themselves head down in the water in a nearly vertical position, these fish are commonly found among the poi-

46 / PARTNERS, GUESTS, AND PARASITES

sonous spines of various sea urchins, where they find protection from predators. The host does not profit from this association.

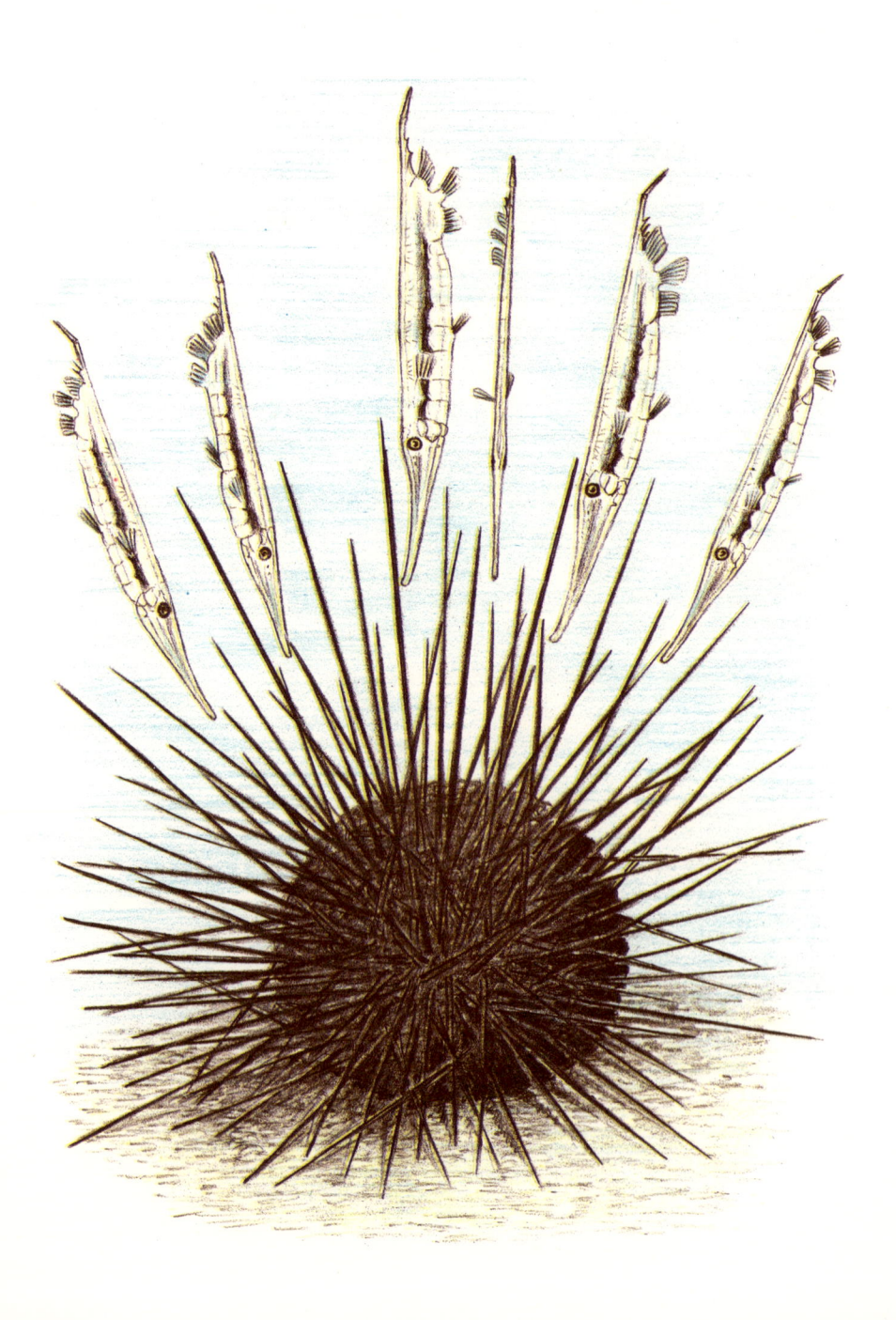

Transporting Passengers

The instances of phoresy—transportation of one partner by the other—noted earlier in this chapter were only matters of convenience and not necessities for those animals, such as the bee eater and the remora, which are able to get around on their own. For other animals, though, it is vital to have a means of transportation to places they cannot reach under their own power.

A great many insects and other creatures, either as larvae or as adults, get their food from the droppings of large mammals, usually grazing animals. Because they will feature prominently as the "working partners" in some of the symbiotic associations we are about to discuss, it will be worth the time to take a closer look at a very interesting group of beetles known popularly as tumblebugs. They belong to the family Scarabaeidae and comprise a subfamily of their own, which includes the most famous member, the sacred scarab of the ancient Egyptians, as well as its other dung-beetle relatives.

The sacred scarab's habits are typical of most other members of its group. A comparatively large beetle measuring a little over an inch long, it is well equipped for its peculiar food-gathering activities. Finding fresh droppings from one of the large grazing animals, the beetle forms a mass of it into a hard ball the size of a large apple and then proceeds to roll this ball away —a seemingly impossible task for so small a creature.

However, insects are comparatively more powerful than most larger animals, and the scarab is a sturdy, chunkily built fellow. Moving backward, it pushes the ball with its hind legs and braces itself with its forelegs. Sometimes two beetles team up together. Upon reaching a suitable spot, the ball of manure is buried, and the beetle, or beetles, then dine leisurely underground without fear of disturbance. When it is all eaten, they come to the surface and search for fresh food.

In the breeding season a mated pair of beetles gather food in the same way for the future offspring. After the ball has been buried, the female begins to shred it, and forms a pear-shaped mass which is connected to the surface by a tunnel through which air can enter. She then lays an egg at the narrow end of the mass, and the hatching larva finds enough food in the soft insides of the manure pear to last until the time for pupating has come.

This scarab is gathering a large ball of dung as food.

A rhinoceros beetle serves as an unwitting "air taxi" for tiny, wingless mite larvae.

Watching the scarab rolling its large ball along, the ancient Egyptians interpreted its activities to fit their ideas of the universe. The beetle represented the sun; the ball, which they believed was rolled from sunrise to sunset, was the earth. As the symbol of a deity, the scarab was represented on Egyptian monuments in carvings and paintings, and stylized scarabs were fashioned from gem stones for use as jewelry.

Scarabs and all the other dung beetles can easily reach fresh manure that is suitable as food. They are strong fliers and simply buzz from one source to another until they find what they want. The situation is quite different and much more complicated for other creatures like mite larvae and worms that also live on such animal droppings. Having no means of transportation of their own, they find themselves in a difficult situation when the manure begins to dry out. They somehow have to reach a new source of dung or else die of desiccation and lack of food. The solution to their dilemma is unique.

As noted earlier, dung beetles really get around in their search for food. Even though mite larvae and worms are blind and cannot see the beetles, they have other senses that announce the presence of the large insects. The mite larvae, for example, smell their prospective transportation with the help of "noses"—olfactory organs—located in their tiny legs. This may sound a bit peculiar to us, but when dealing with insects and lower animals, we must get rid of the preconceived notion that sensory organs have to be located in the head. After all, crickets and other members of that group hear with organs in their legs and butterflies taste with sensors in their feet.

In any case the mite larvae's legs do tell them when a suitable air taxi has arrived, and by instinct they know that the taxi will probably take them to a good fresh food supply. So they scramble aboard and hold on tight. Soon they are air-borne, and before long a successful landing on a new and, to the mite larvae, delightfully moist and fresh mass of dung has been skillfully accomplished. Off come the passengers, and the beetle goes on its way, perhaps not even aware of, and certainly not bothered by, the fact that it has been used as a fast and efficient means of transportation by nonpaying passengers.

In the case of the worms, the situation is very similar, except that they find their air taxi by touch. In order to accomplish this, they stand up on their rear ends and wave about in the air until they have touched what, here again, instinct tells them is the right vehicle of transportation. Once they have located a beetle, they quickly get on board, and make sure that they will not fall off during flight. Some of these worms secrete a liquid which hardens when exposed to air. With the help of this secretion, they glue themselves to the beetle's legs. Upon reaching a new source of dung, the moisture softens the glue, and the worms disengage themselves and leave their taxi.

Other worms, in order to escape the drying-out effects of the air en route, make their way to the beetle's abdomen and hide under the wings. This led to the erroneous idea, years ago, that these worms are parasitic on the beetles. It took long and close observation to dispel this assumption and ascertain the facts, which are that the worms want nothing from the beetles except transportation, and are in a hurry to get off when a new source of food has been reached.

Not all stowaways that hitch rides on flying insects are as harmless as the mite larvae and the worms which associate with dung beetles. There are a number of small insects which have extremely sinister motives in selecting certain larger insects as means of transportation.

There is, for example, a tiny beetle which will lie in wait for bumblebees as the latter search for nectar in flowers. When a bumblebee lands on a blossom in which one of these female beetles is hiding, the unwelcome passenger will clamp its mandibles firmly around

A bumblebee transports a small beetle into her nest, where the beetle larvae will later feed on the bumblebees' brood.

the bee's "tongue" as it is extended to lap up the flower nectar. The bee cannot get rid of the hitchhiker, and carries the beetle along to its underground nest. There the passenger disembarks, only to lay her eggs in the bees' brood cells. The hatching beetle larvae feed on their host's brood, pupate, and finally leave the bumblebee nest as adult beetles to find mates and start the cycle all over again.

Other beetle species parasitize solitary bees. These bees do not live in communities that have a queen and workers to feed the larvae. Instead, each female bee makes a nest, stocks it well with food, and then deposits an egg in each cell.

Bees of this type fall victim to the larvae of certain species of the oil-beetle family, of which the well-known Spanish fly is a member. Soon after hatching, the minute larvae of these oil beetles climb up on

Guests, Passengers, and Parasites / 53

flowers and wait for bees to come along. However, at this stage it is still a lottery: the larvae have no way of telling which flower-visiting insect happens to be the host species. They will attach themselves to any insect that comes along. If it is a fly, a wasp, or even the wrong kind of bee, such as a honeybee, the larva is out of luck and will die. If, however, the right kind of solitary bee visits the flower, the larva's worries are over. Crawling on the bee's body, it holds on tight to the furlike hairs while the bee is preparing and pro-

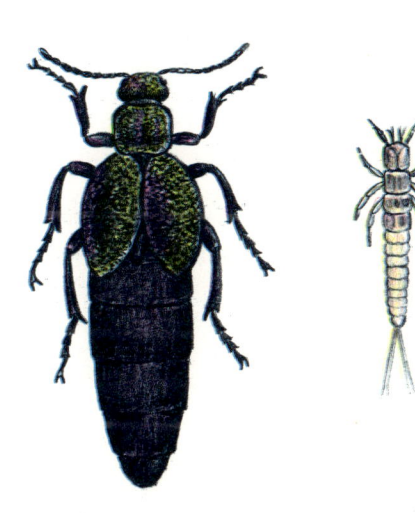

The oil beetle (left) and its tiny first-stage larva which parasitizes solitary bees.

visioning her nest. Then, when she lays an egg in one of the cells, the larva lets itself slip down on the egg, and is sealed into the cell by the bee. The larva promptly eats the egg, thus disposing of a competitor, and then dines in leisure on the honey, which is sufficient to last it through the larval stage.

Unwelcome Guests

All social insects are beset by uninvited guests that take, but give nothing in exchange, and generally make a nuisance of themselves. There are, for example, certain moth caterpillars which invade beehives. They are mostly scavengers, although they will occasionally eat some bee larvae also. The worst damage they do is to chew through the wax cells and to entangle the hive with silken threads.

The bee louse is a different kind of nuisance. These minute insects, though commonly called lice, are really flies. The bee louse clings to the bee's furlike hairs, and steals food intended for others. The little thief has an ingenious method: knowing instinctively that honeybees are conditioned to give up a drop of honey when tapped on the mouth parts by a bee larva, the bee louse simply imitates the grub, taps the bee on the mouth parts, and is rewarded with a bit of honey really intended for the bee's own brood!

Even more than bees, ants have to put up with such dubious guests. Some of the guests have to step lively to keep from being caught by the ants, which obviously resent them. Others, being mainly scavengers—such as the larvae of certain flower flies—are usually tolerated by the ants, and are in any case protected from attack by their thick, leathery outer skin.

Stealing food as it is passed from one worker ant to another, the silverfish is an unwelcome guest in an ants' nest.

Feeding on the body secretions of its host, this small wingless cricket has to step lively to avoid being killed by the ants.

Almost every type of insect from primitive silverfish to crickets, beetles, and even parasitic ant species are found living in ant colonies. Collectively, ants are such popular hosts that several thousand different species of myrmecophiles (literally, "ant lovers") have chosen ant communities as their permanent abode. They are often referred to as "guests," an especially inappropriate and misleading term in most cases, for far from being guests in our meaning of the word, they are almost always uninvited and usually unwelcome. At best, the majority are like "the man who came to dinner," as one biologist put it. At worst, they are predatory and harm the community by attacking and eating the ant brood. Some do fit the description of guest, or rather subtenant: tolerated or even welcomed by the ants, they render services in return for food and shelter. The majority, however, are truly parasitic and usually get their food by stealing it.

Many tiny creatures—primitive silverfish, springtails, and mites—steal food as it is passed from one ant to another. One silverfish even mimics an ant grub and manages to fool the ants into feeding it just as they would one of their own grubs!

One of the most interesting of these minute creatures

Symmetrically balanced on the body of a worker ant, these tiny mites live on the food they steal from their hosts.

is a species of mite so tiny that a number of them can ride on a single ant without bothering their host. They also are on the lookout for a chance to snatch a bit of food either from their host ant or from another that comes close. The really fascinating fact about these mites is their technique of arranging themselves on their host. Regardless of how many ride on a single ant, they always arrange themselves in a symmetrical pattern in order to keep the load balanced. If three mites ride on one ant, the first will occupy the choice spot, which is on the head, and the other two will place themselves, respectively, on the right and left sides of the body.

A tiny fly larva has still another way of getting food from the ants. By means of a special suction disk, it clings firmly to the neck of an ant larva. As the larva is being fed by one of the worker ants, the fly grub simply reaches over and steals part of the ant larva's food. The ants do not seem to notice anything untoward, and will go on feeding both their own brood and, unwittingly of course, their uninvited guest.

In addition to all the unwelcome and usually semi-parasitic invaders of ant communities, there are also many that are welcomed by the ants because they give something in return for what they take or are given. Despite the fact that some of these guests prey upon the ant brood and are thus actually harmful to the community, there is an element of mutualism even in such relationships, for the ants do receive at least some compensation, usually in the form of sweet or otherwise relished secretions. We shall hear more about a few of these interesting symbiotic associations in the chapters dealing with the mutually beneficial partnerships.

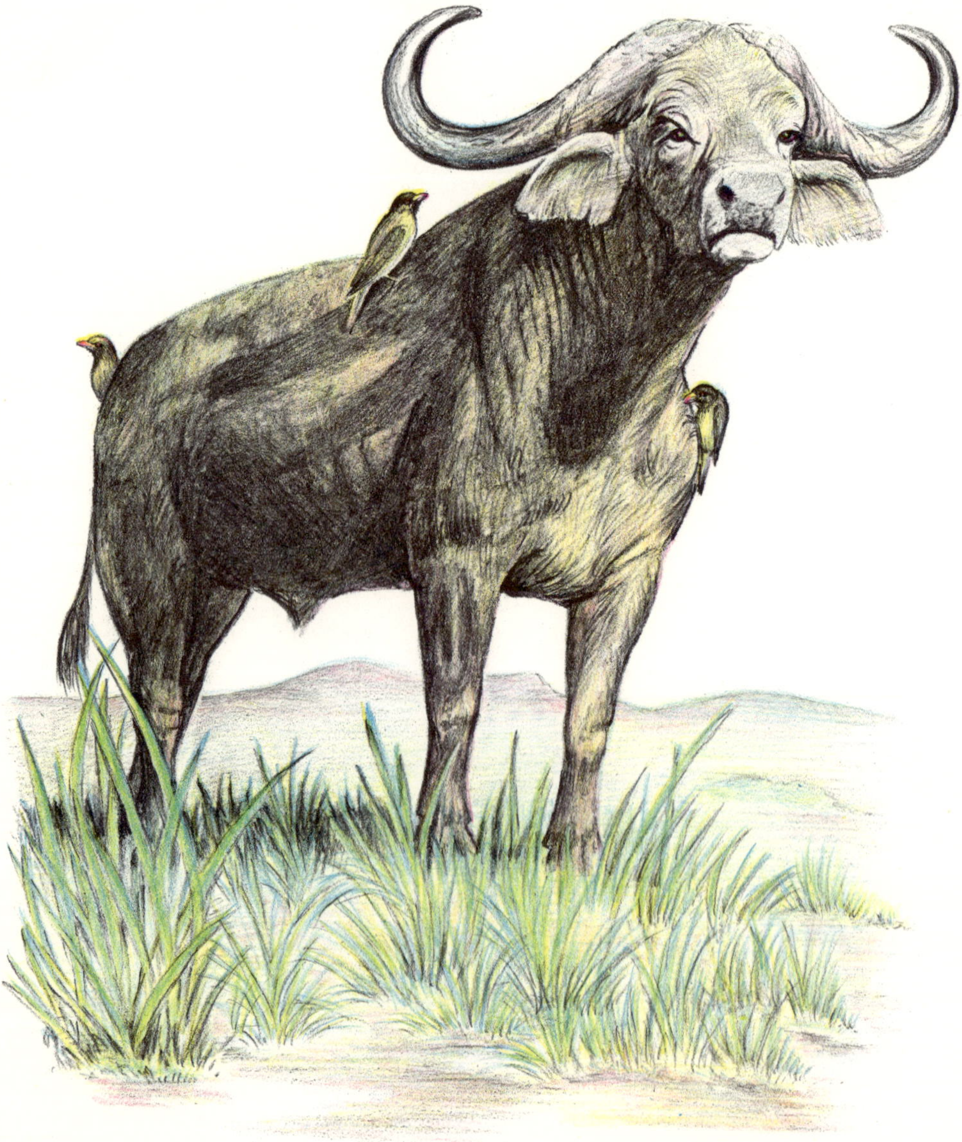

Partners for Gain and Convenience

Symbiotic relationships offering benefits to only one side are obviously not partnerships. As we have seen in the preceding chapter, most of the so-called partners that gain from the associations are in reality uninvited guests that exploit their companion or his activities for their own ends. In some cases, the beneficiaries are even outright thieves and social parasites that will take from their hosts but give nothing in return.

Any true partnership demands that both sides profit. In nature such mutually advantageous arrangements are the ideal type of symbiotic association. Called *mutualism* by biologists, this kind of symbiosis is widespread among many different groups of animals on the land as well as in the water. Some of these true natural partnerships are formed to obtain some benefit or convenience essential to existence, while other partnerships are vital to the good health or even the survival of one or both partners.

It is often assumed, mainly because many textbooks on biology convey this impression, that every action by animals is in some way related to their survival as an individual or a species. This approach, usually couched in dry scientific terms, casts a gray and humorless shadow over the study of animals. If every single movement, every action by every animal has to be seen in the context of its struggle for existence, most of the fun of studying animals is lost. But watching a kitten or a puppy playing makes it obvious that animals do at least some things just for the fun of it. We do not have to ascribe human thoughts, characteristics, and reasoning to animals to realize that many of their actions can be explained only by a zest for life and a heightened sense of well-being. Many wild animals, including adults of different groups, display any number of what we somewhat arrogantly term "human" inclinations. The bear that becomes temporarily oblivious to its surroundings when it finds the honey it relishes so much, the otters that squeal with sheer pleasure as they take turns going down the mud or snow slides they build for themselves, and the bower bird that so painstakingly paints and decorates its bower may only be following their instincts. But no one can doubt that they are enjoying themselves very much while doing it.

Hence it should not come as a surprise that many of the mutually beneficial symbioses do not involve anything even remotely related to the survival of the animals concerned. In many cases it is the pursuit of a luxury or delicacy that motivates these relationships. We find that creatures with entirely different ways of life team up temporarily in order to gratify their desire for obtaining some delicacy.

In one such team of a bird and a mammal, neither animal needs the food they seek together in order to stay healthy and survive, but they evidently enjoy it very much. And so the association is successfully formed, the partners share the spoils, and afterward each again goes its own way. The partners involved in this relationship are the badger-like ratel of Africa and the honey guide, a rather drab-colored bird looking somewhat like an overgrown sparrow. Because of its unusual and eye-catching features, the association was noted centuries ago as a natural curiosity. Since then it has often been recounted, sometimes with fanciful embellishments.

The ratel is an Afro-Indian member of the weasel family. Superficially it resembles the North American skunk, being about the same size, black below but solidly ash gray above from head to tail, without the two white stripes that distinguish the skunk.

Ordinarily the ratel eats very much the same food as its close relative the badger. Omnivorous, it will take anything from roots and berries to insects, bird eggs, and young birds, as well as small reptiles and rodents. However the ratel also has a "sweet tooth," which makes it inordinately fond of honey and eager to obtain it whenever possible.

In many other parts of the world, the ratel would be more or less out of luck, for it cannot climb well and therefore could not reach bee nests that are built high up or in trees. Most African subspecies of honeybees, however, build their nests underground in deserted rodent burrows or in old termite mounds. Whenever the ratel finds a hive, it tears up the ground or the termite structure with its strong, sharp claws, exposes the combs, and feasts on the honey and the fat juicy

Waiting for the ratel to catch up, the honey guide leads its companion to a wild bees' nest.

grubs. The attack of the enraged bees bothers the ratel very little, for in addition to its thick, heavy fur, it has a layer of fat beneath the skin that renders the stings of the bees ineffective.

The ratel's liking for honey has been well known to the inhabitants of South and Central Africa for centuries. Indeed, the word "ratel" is derived from the Dutch word *rateldaas,* meaning "honeycomb badger." However, the instinctive knowledge of the ratel's "sweet tooth" by its avian partner, the honey guide, goes back a much longer time to the nebulous evolutionary beginnings of this unusual association.

Despite its sparrow-like appearance, the honey guide is not related to the sparrows. Zoologists classify it and its relatives in a family whose scientific name is Indicatoridae—a most appropriate term, because in Latin it means "those that point out."

Normally the honey guide feeds almost exclusively on flying insects and especially on stinging species from which most other birds will shy away. The stings do not seem to bother the honey guide at all. It will snap up bees in the manner of a flycatcher capturing flies on the wing.

Its unusually thick, leathery skin undoubtedly helps protect the bird against the stings of the insects on which it commonly preys. Another distinctive anatomical feature is its feet, which like those of woodpeckers have two toes pointing forward and two backward.

The nesting habits of the honey guide explain why it was popularly called "Honigkuckuck"—honey cuckoo —by the German settlers in the former German colonies of South Africa: it does not build a nest, nor does it raise its young. Like the cuckoo, the female deposits her egg in the nest of another, often closely related bird. Usually she destroys the eggs she finds in the nest. After hatching, the young honey guide commits an even more unpleasant act. With the help of a special calcified projection on its beak, it fatally bites its legitimate nest-mates, so that it alone is left to get all the care and attention of its foster parents. This habit, which is of course an instinctive action designed to insure the survival of its species, is shared with the European cuckoo, whose young also kill their foster brothers and sisters. Obviously the population of cuckoos and honey guides has to be relatively small in proportion to the foster species, or the latter would soon be exterminated.

Although their nesting habits are interesting, if unattractive, the unique behavior displayed by the honey guides in leading mammalian partners to beehives has become almost legendary. It is not surprising that many observers in the past century have invested the honey guides with near-human powers of reasoning and planning.

As soon as the bird has located a honeybees' nest, it flies off in search of a partner. Usually, the bird finds a ratel, but frequently—and this is quite amazing—the honey guide selects another animal such as a baboon or even a man as its companion in the pursuit of the food it desires. How the bird knows which animals like honey is still a mystery.

As soon as it finds its prospective partner, the honey guide utters loud churring sounds and flies ahead of the animal, repeating its calls from time to time and waiting patiently until its teammate has caught up. After finally arriving at the site of the beehive, the bird sits quietly in a nearby bush and waits until after the nest has been uncovered and the partner has eaten its fill.

It was assumed in the past that the bird was interested mostly in the bee grubs and pupae, and perhaps even to some extent in the honey. However, this has been proven erroneous by more recent research. Amazingly enough, it is primarily the wax of the combs which the honey guide covets. Why it would want to eat this substance of low nutritional value is still unknown, but it has been determined that the intestinal tract of this bird harbors certain microorganisms which help it digest the wax. This means that the bird has to live in a double symbiosis with both the ratel or other mammals, and the microorganisms in order to obtain and utilize the wax it likes to eat.

To the natives in those parts of Africa where the honey guide is found, the service rendered by the bird in leading them to honey is much appreciated. Because it is so unusual a phenomenon, a host of legends and superstitions have sprung up in connection with it. Many Africans firmly believe that the bird will punish them if it helps them locate a bees' nest and then does not get its fair share. They will never harm these birds in any way, being convinced that the honey guides can revenge themselves by bringing bad luck or leading them to a lion's den the next time. For this reason, naturalists who wanted to obtain a specimen for study found it all but impossible to get any help from the natives, and were in fact met with open hostility when they were observed shooting one of these birds.

Despite the fact that we call the honey guide's actions "instinctive"—a word that covers a great deal about which we know very little—there are many puzzling questions raised by this bird's behavior. How, for instance, do the birds know which mammal will respond to their invitation to search for honey? Or do they simply try any large animal they find and wait to see whether or not it shows interest? Future research may throw light on at least some of these questions.

One of the best-known examples of symbiosis, and

The Egyptian plover, also called the crocodile bird.

certainly the one that was recorded earlier than any of the others, is the association between the African crocodile and the small, attractive Egyptian plover, or trochilus, commonly called the crocodile bird. First reported almost 2500 years ago by the Greek historian Herodotus, the relationship was described by the Roman author Pliny the Elder in an entertaining account of what he calls the "friendship" between the large reptile and the small bird:

> . . . As the crocodile lies resting on the dry land with wide-open mouth, trochilus flies over, creeps into its mouth, and cleans it. This is pleasing to the crocodile, and it therefore spares the bird, to the extent that it even opens its mouth wider so the bird has no difficulty when it wants to come out again.
>
> This bird is small, not larger than a thrush, stays near the water, and warns the crocodile against the mongoose by flying over and waking it, partly with its voice, partly by pecking it on the snout.

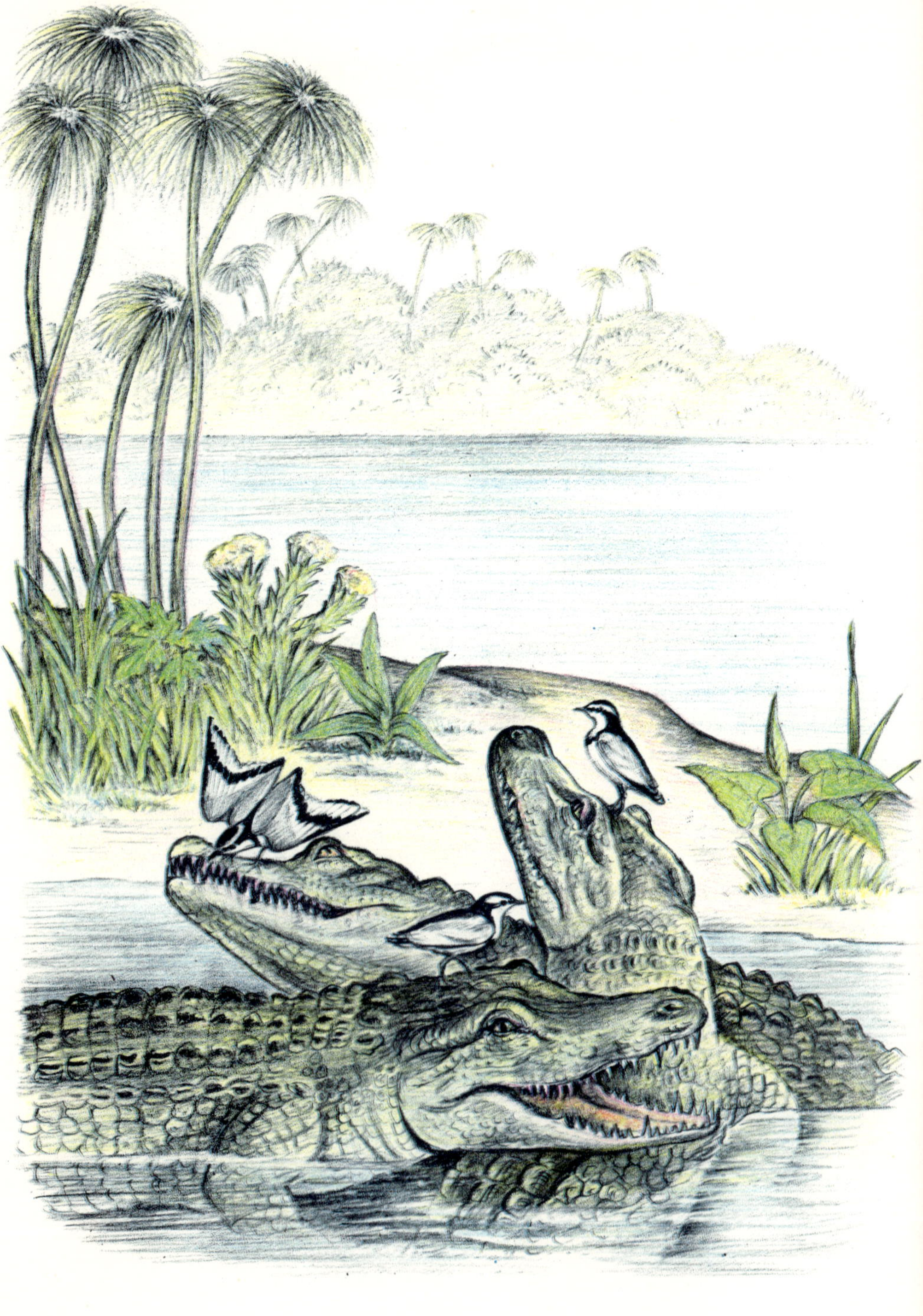

While this description is largely quite accurate, parts of it are altogether erroneous, as for instance, the bird's warning the crocodile against the mongoose—a misconception based on an ancient fable that originated with the Egyptians and was adopted by the Greeks and Romans.

The African mongoose—which is larger than its Indian relative, the famous Rikki-tikki-tavi of Kipling's *Jungle Book*—was held sacred by the ancient Egyptians, who considered this so-called Pharaoh's rat the deadliest enemy of the crocodile. According to the legend, the mongoose would stealthily approach a sleeping crocodile, jump into its wide-open mouth, and work its way down the throat and into the chest cavity, where it would eat the reptile's heart. Finally, covered with blood, the courageous beast would emerge triumphantly from a hole it had torn in the dead crocodile's belly!

Needless to say, no part of this fable is true. Mongooses, which are related to the civet cats, are great predators for their size and have become famous for their battles with poisonous snakes. It is quite possible that the African mongoose eats crocodile eggs or the newly hatched young if it can capture them without danger from the mother, which often guards the eggs. That, however, would be the extent of the mongoose's threat to crocodiles.

As for the plover, it is a fact that this bird—which is not a true plover but a closely related courser—feeds on various parasites from the crocodile's skin, and cleaning the big reptile's teeth is only an extension of this activity. Though long doubted by zoologists, the role of the bird has been verified by several eye-witness reports of reliable observers. The likelihood that the bird does indeed walk around fearlessly in the crocodile's open mouth is further strengthened by the discovery of many similar cleaning activities among other animal groups, about which we will hear more later.

It also is a fact that this attractively patterned black, white, and gray little bird, which stands only about twelve inches tall, serves as a sentinel, warning the crocodile, though not of an approaching mongoose, but certainly of other large creatures. Its shrill, loud cries immediately alert the shy reptile.

There is no evidence that the crocodile ever snaps its mouth shut on its "living toothbrush." Cynics have suggested that the reason for this is the extreme unpalatability of the plover, but crocodiles are not very choosy about their food. In any case, it is unlikely that this explanation can account wholly for the apparent immunity of the plover.

Some truly amazing symbiotic relationships between animals that would normally be antagonistic to each other are found in the ocean, especially those associations involving sea anemones. These creatures are primitive animals of the order Actiniaria. Because of their shape, their colors, and the tentacles that grow in large numbers around their mouth openings, they bear a superficial resemblance to flowers.

Partners for Gain and Convenience / 71

Among the many associations between fish and sea anemones we find all degrees of mutualism, from a comparatively loose voluntary relationship to an almost complete dependency upon each other. These latter instances will be discussed in the next chapter.

Sea anemones live mainly in warm, shallow waters, although a few are found in cool regions. Generally the anemone has a short cylindrical body and a flat pedal disk by which it attaches itself to solid objects such as rocks and shells. The slitlike mouth opening is connected to the stomach cavity by a gullet. Around the mouth are grouped the hollow tentacles on which stinging organs called nematocysts are located. The stinging organs are minute rounded capsules that can be everted as a help in capturing prey. A small process on each capsule, evidently stimulated by certain chemical substances, triggers the discharge of a thread tube, normally coiled inside the capsule, which bears at its base several spines and thorns. Capable of penetrating the skins of animals, this tube injects a fluid called hypnotoxin that can paralyze small animals and causes a painful stinging, burning sensation in larger ones.

A nematocyst, or stinging organ, before (left) and after being discharged.

A sea anemone eats various marine creatures, including small crustaceans, mollusks, and fishes, which are paralyzed by its tentacles and carried to its mouth. Undigested parts are ejected via the mouth opening.

By and large, coming within reach of the tentacles means death to any small fish, which are generally well aware of this fact and avoid swimming too close to the beautiful but dangerous anemone. However, certain small fish not only venture close, but actually more or less live among the deadly tentacles without being harmed by the anemone—a phenomenon that in the past has puzzled observers.

Most of these fish have brilliant and striking color patterns. Take, for example, the bright orange, white-and-black-banded clownfish, *Radianthus ritteri*, frequently found among the tentacles of large sea anemones. There is no doubt that the fish gains considerable protection from the association, for predators that would consider it a tasty morsel keep a respectful distance away from the anemone. It is not quite as clear how the sea anemone benefits from the presence of the clownfish; however, marine biologists commonly believe that it lures other fish within reach of the tentacles, and thus helps the anemone to obtain food. When the clownfish darts out to catch smaller prey, the anemone frequently gets its share after the fish returns.

The most intriguing question, of course, is why these fish enjoy immunity from attack by the anemone's deadly weapons. Recent studies have thrown some light on this mystery. In the past, it was believed that a fish

Partners for Gain and Convenience / 73

had to touch the tentacles in order to trigger the discharge of the thread tubes. This, however, has been found not to be true. Instead it seems that certain chemical substances trigger the action, a fact that was proved by pouring these chemicals into the water.

Brightly colored clownfish live unharmed among the poisonous tentacles of a sea anemone.

Close observation of the symbiotic clownfish disclosed that they secrete a slimy fluid when they come close to and swim among the tentacles of the anemone. The chemicals in this mucus apparently inhibit the discharge of the nematocysts: the anemone does not "recognize" the fish as something edible. However, each of the symbiotic fish associates with only certain species of anemone, and its immunity does not extend to others.

Another interesting partnership exists between sea anemones and crabs in a relationship that in some cases demands a considerable amount of cooperation by both partners in order to set up proper housekeeping. The crabs, mostly of the type known as hermit crabs, which protect their soft bodies by occupying empty snail shells, are the ones that initiate the action and also usually receive the greater benefit from the partnership. Watching a crab as it proceeds to find itself a partner is quite entertaining, so we will follow *Pagurus arrosor*, a common species of hermit crab, as it goes about its task.

Picking out a sea anemone that is to its liking is not very difficult. Now, however, comes the job of persuading the anemone to abandon its home on the rock to which it is attached and transfer itself to the shell which is the crab's portable house. The move is not made without considerable coaxing by the crab, for the anemone's pedal disk is firmly attached to whatever surface it is sitting on, and it cannot simply be hoisted onto the shell. Instead the crab approaches its prospective partner and begins to tap and stroke it with its feelers. This activity proves to be persuasive, for the anemone begins to detach itself from its base and finally, after a little more coaxing, lets go altogether. Now the crab actually embraces the anemone, lays it on its side, and then manipulates it onto the back of the shell. Once the anemone feels the shell's surface with its pedal disk, it usually attaches itself to it without further delay. *Pagurus* straightens up and marches off carrying its life insurance on its back.

The presence of the anemone does give considerable protection to the crab. One scientist observed an octopus that attempted to attack one of these crabs but was successfully beaten back by the stinging tentacles of the crab's formidable partner.

This hermit crab has two sea anemones attached to its shell.

While carrying around a house with "built-in" life insurance has great advantages for the crab, it also creates problems. The crab molts and grows, and the day inevitably comes when it has to move into a larger house. But what about the anemone which has anchored itself firmly to the old shell, thereby acquiring free transportation? It probably would be simpler to just let it be and to look for a new partner along with a new house. However, the crab does not do that; apparently it has grown accustomed to the anemone. Again using the same technique of stroking and patting, the crab persuades the anemone to let go of the old shell and attach itself to the new one.

Sea anemones are not the only marine creatures with stinging organs that are found in close association with other creatures. Certain large jellyfish known as the Portuguese man-of-war, whose long trailing tentacles with their millions of nematocysts can deliver extremely painful stings even to human beings, have established similar symbiotic relationships.

Like sea anemones, jellyfish eat any small animal they can reach and overpower with their tentacles—with one notable exception: a small pretty fish with the long name *Nomeus gronovii gmelin*. Commonly known as the man-of-war fish, it lives among the tentacles of *Physalia arethusa*, the large jellyfish often found in the Gulf of Mexico.

The small fish are conspicuously patterned in light blue-gray and black stripes. Swimming fearlessly among the jellyfish's tentacles, they dart out to grab

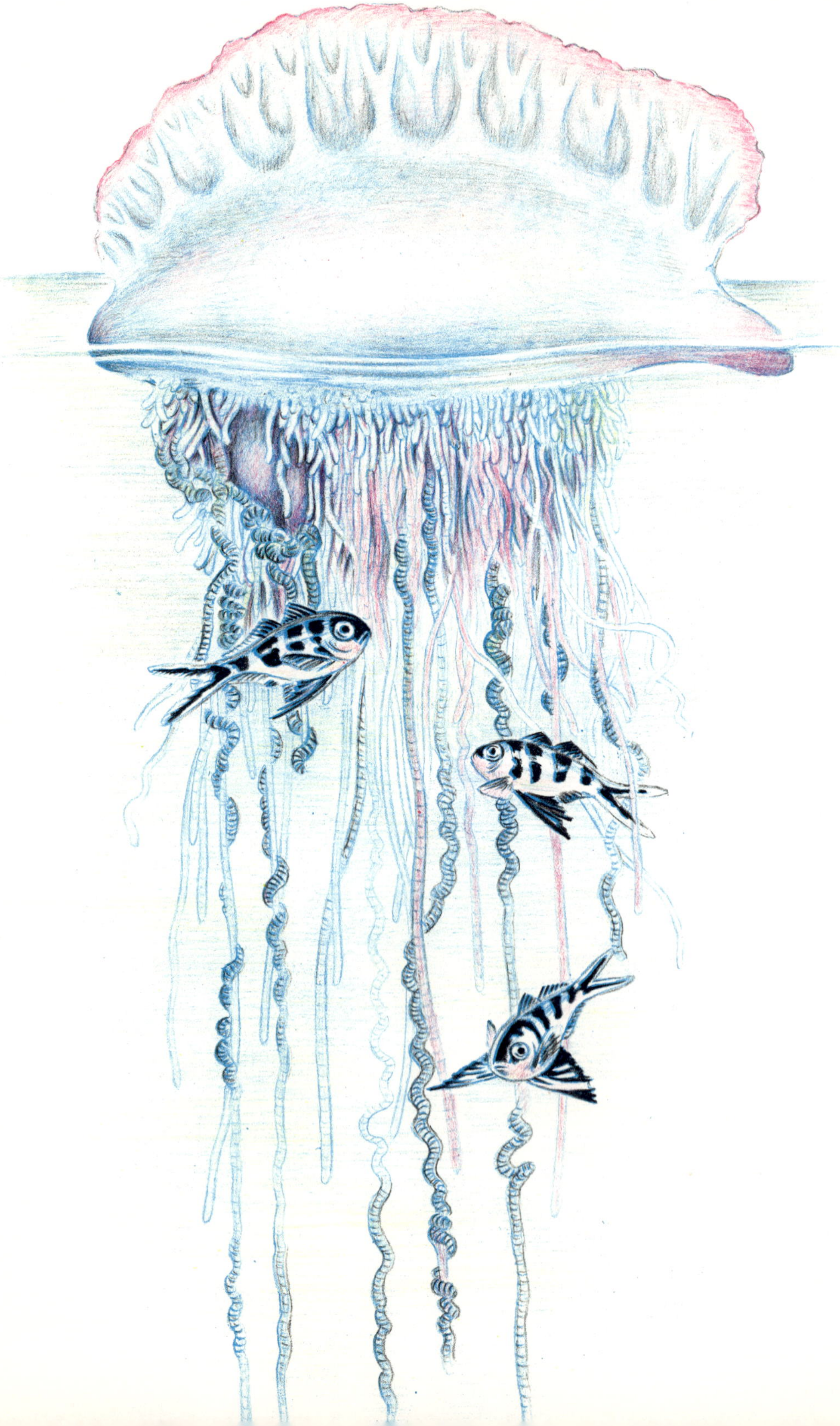

some small prey and then return to the protection of their large dangerous partner. As in the case of the anemone and its symbiotic fish, the Portuguese man-of-war undoubtedly profits by the presence of its small partners, which act as bait for other creatures the jellyfish catches and eats as they venture too close to the tentacles in pursuit of *Nomeus*.

The hot, humid regions of the tropics vie with the ocean in their abundance of animal life. Nowhere else do we find so many insects, both attractive and unattractive. The biting, stinging, and sucking species—all types of parasites—are more numerous in the latitudes near the equator than anywhere else on earth.

Most animals in the tropical regions, including man, suffer from infestation with one or more kinds of parasites. It is well known that serious health hazards frequently result from such infestations. Less well known is the fact that this predicament is shared by many of the large insects of those regions, which often are parasitized by a variety of small creatures, especially mites.

Tiny eight-legged members of the class Arachnoidea, the spiders and spider-like creatures, mites are often found as parasites of both plants and animals. The great majority of the animal parasites live on the skin, blood, and other tissues and body fluids of land vertebrates. The species we are discussing here, however, have selected insects as their hosts and victims.

One of the largest and most spectacular of all tropical

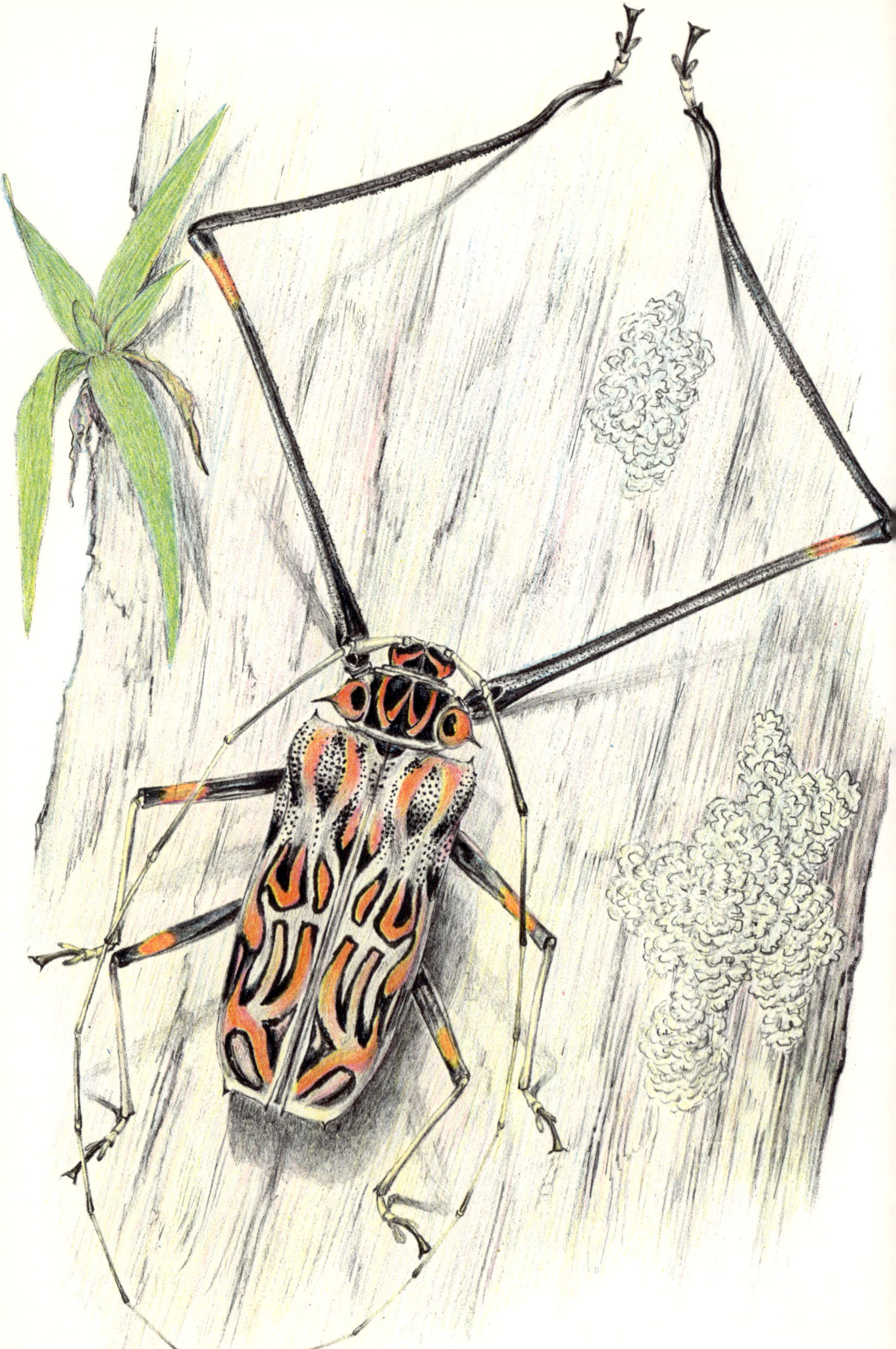

The small pseudoscorpion, which here is shown greatly enlarged, is frequently found riding under the wing covers of the harlequin beetle.

insects is the harlequin beetle. It belongs to the group known as longhorn, or longicorn, beetles. Strikingly colored in patterns of bright orange-red, black, and beige, its four-inch body length is dwarfed by forelegs that are twice as long, and antennae measuring some six inches. When the wings are fully spread, they measure more than ten inches across.

The portion of the beetle's anatomy most vulnerable to the attacks of parasites is the soft abdomen. The cracks along the segments permit the mites to insert their sucking mouth parts to feed on the beetle's body juices. The insect has no way of getting rid of its unwelcome guests. However, help in the shape of another member of the spider group is frequently available.

Pseudoscorpions look like miniature scorpions minus the tail. Unlike scorpions they are completely harmless, as they possess no poison and neither sting nor bite.

82 / PARTNERS, GUESTS, AND PARASITES

They usually live unnoticed under stones, bark, or moss. Some make their homes in books or furniture. All feed on minute creatures such as various types of mites.

Luckily for the harlequin beetle, some pseudoscorpions do not mind exchanging their natural habitat for temporary living quarters on the beetle's body. There, they get busy on the mites that have established themselves in the cracks between the abdominal segments. After a while, the pseudoscorpion succeeds in ridding its host of the parasites. It then leaves for other hunting grounds. Both the beetle and the scorpion profit from this intimate, if temporary, relationship. While this case has not yet been studied closely enough to determine just how serious a health hazard the unchecked infestation would be to the beetle, there can be no doubt that the pseudoscorpion performs a valuable service for its partner, while assuring itself of a convenient and bountiful food supply during its association with the large insect.

An enlarged portion of the harlequin beetle's abdomen shows pseudoscorpion feeding on the mites that infest the large insect.

The fact that pseudoscorpions helped to rid large insects of mites was discovered in 1892 by a German zoologist. This, incidentally, was the first time a naturalist drew attention to a symbiotic association involving the cleaning of one partner by the other. We shall hear more about this phenomenon in the next chapter.

Another type of cleaning symbiosis is found among ants, involving certain tiny "guest" ants, *Leptothorax provancheri*, commonly called barber ants, which are the most independent of all the creatures that live in ant nests. While they maintain their own separate brood chambers in the nest of *Myrmica brevinodis*, a common North American species of larger ants, and do not permit their hosts to enter these chambers, they are welcome guests all the same. The reason is that they climb on the backs of their hosts and groom them, a habit which earns them their common name. This apparently pleases the larger ants very much, for they try to encourage their "barbers" by feeding them. This is an example of a highly successful association in which a valuable service is exchanged for food, shelter, and protection.

It is common knowledge that ants relish sweet substances and that their truly phenomenal sense of smell quickly brings them in droves to wherever the sweet stuff is located. Anyone who has ever gone on a picnic and any housewife in the suburbs who has ever spilled a little jelly on the floor will readily concede that ants have an extraordinarily keen sense of smell.

84 / PARTNERS, GUESTS, AND PARASITES

Ants vigorously pursue the sweet substances they relish, such as sugar and the secretions of certain insects they evidently consider a delicacy. Ants the world over are known to have their own "livestock" which they milk the way we milk our cows: the various species of aphids, or plant lice, give off a sweet excrement whenever the ants stroke them with their feelers. In the hollow stems of certain South American plants, ants and beetles fight pitched battles over species of scale insects whose secretions they both value so much that each wants to keep these "cattle" for themselves.

In order to get the most out of their herds of aphids, ants go to great lengths to protect and care for them. Some ant species build regular "stables" for their aphids and attend them solicitously. Others transport their "livestock" to better feeding grounds—thereby frequently causing great harm to cultivated plants—and still others carry the aphids underground when winter comes, to protect them from the cold weather.

"Milking," ant style: an ant persuades an aphid, by stroking it with its feelers, to give up a droplet of sweet liquid.

All ants defend their aphids against enemies. In exchange, aphids give off droplets of "honeydew" whenever their protectors "milk" them by stroking and patting.

Aphids, however, are not the only insects that provide ants with delicacies. Some other insect guests are highly welcome in the ants' nest because they too have a secretion that is relished by their hosts. Among them are several species of beetles; other species of the same beetle family live as guests in termite nests. The beetle *Lomechusa* stays with two different species of ants, changing quarters after it has completed its larval stage.

The beetles and their larvae are treated in every respect like members of the ant community; they are fed, groomed, and protected. This is all the more surprising because the beetles eat some of the ant brood. The ants seem either not to notice or not to care. They frequently give so much attention to the beetle brood that they neglect their own—and all this just for a chance to nibble at the special hairs which secrete the substance they are so keen to obtain.

Luckily, perhaps, the care provided for the beetle larvae proves to be too much of a good thing in many cases. Beetle pupae are not encased in cocoons like those of ants, and are thus more susceptible to damage by the constant turning and moving which, though necessary for proper development of the ant pupae, is harmful to that of the beetles. Many die in the pupal stage, which prevents large populations of these beetles from infesting the ant nests and killing a sizable part of the ant brood.

Many beetles that live as guests in ant nests have evolved special glands giving off secretions relished by their hosts, but most of them have become dependent upon the ants and could not live outside of the ant nest.

Among the strangest, and in many cases, most complicated of all relationships found between ants and other insects are those involving a family of mostly small butterflies called blues, found in all parts of the world. Not all of them are blue; many are brownish or whitish, and in some species, the males are blue while the females are brown. The majority are delicate-looking butterflies with a wingspread that rarely attains two inches.

The habits of these blues are much more interesting than those of most other butterflies. The caterpillars, which are sluglike in appearance, in many cases have honeydew glands on their backs. These species have evolved symbiotic relationships with ants that vary greatly in degree from independent mutualism to complete dependency. Most frequently, however, the association is apparently mutually beneficial, but more or less voluntary. As the caterpillars feed on plants, they

A common species of blue butterfly, showing upper and lower wing surface.

are visited by ants, which induce them to give off a droplet of honeydew by stroking them in a manner similar to that used with aphids. In exchange, the ants undoubtedly protect the caterpillars against a variety of enemies. The larvae of some species invade the ants' nests, but are welcome because of the sweet secretions with which they "repay" their hosts.

How far mutually beneficial symbiotic relationships have progressed in many cases, how highly specialized and complicated some have become in order to permit the partners to derive maximum benefits, and how greatly other creatures are often affected by such "partnerships for survival," are discussed in the next chapter.

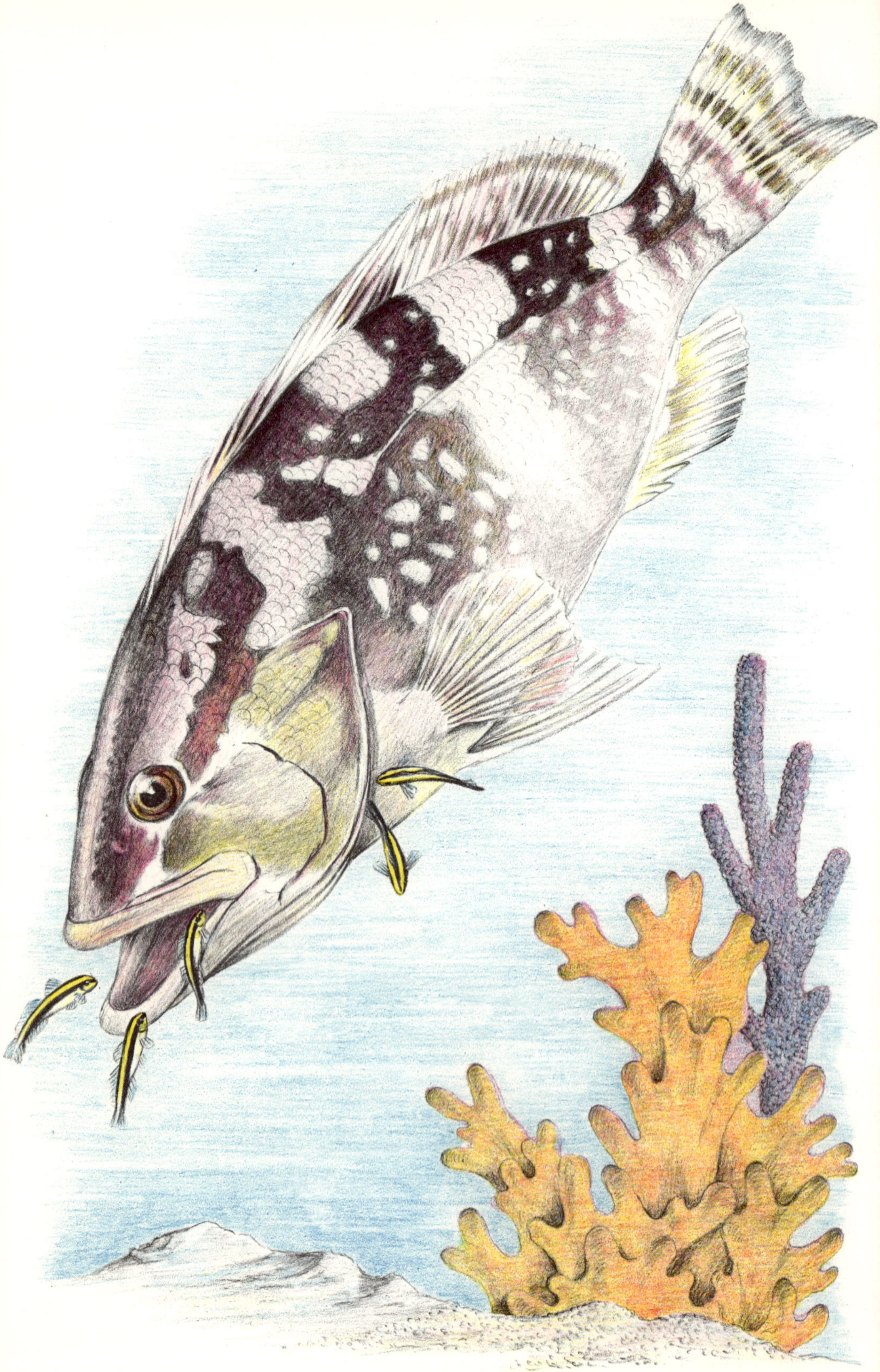

Cooperation for Survival

In the struggle for existence all organisms strive to utilize anything that will give them a better chance to assure their survival. For many creatures, teaming up with other organisms—either plant or animal—in a mutually beneficial relationship has proven very successful. Moreover, a great number of plants and animals have evolved such highly specialized associations that the survival of one or both partners has become entirely dependent upon continuation of their intimate relationship. This is mutualism at its peak. Many of these associations offer convincing evidence that certain symbioses between two widely differing organisms may, by assuring the survival of the partners, have far-reaching effects upon the entire environment.

The best known and, indeed, classic example of symbiosis between plants and animals essential to at least one of the partners, but also affecting many other creatures, is the close relationship between nectar-seeking insects, especially honeybees, and flowering plants.

There are many other much more specialized associations between specific plants and particular insect species. But there are no symbiotic associations involving mammals that are true dependencies. As we have seen in the preceding chapters, mammals figure mainly in one-sided associations as the partner whose activities are exploited by the other one for the latter's own benefit. Only in comparatively few cases, such as that of the ratel and the honey guide, or the cattle and the tickbirds, do the mammalian partners also reap some benefits from the association. Generally it is the animal lower in the evolutionary scale that profits.

This is not really surprising. Mammals are the most advanced and the most highly evolved group among all animals on earth. They have a more highly developed brain and nervous system, and superior methods of reproduction and rearing of young. All this gives them more freedom and adaptability in their way of life, and therefore reduces the necessity of having to depend for survival upon alliances with other organisms—always excepting, of course, their food sources, both plant and animal, and the various microorganisms that are essential to health. No mammal, however, is a parasite, and very few form symbiotic relationships with other creatures that contribute directly to their chances of survival; the few that exist are loose associations which never develop into dependencies, but which are interesting because they involve a pooling of the senses of sight and smell by two very different animals.

The senses among the two highest groups of animals,

birds and mammals, are not equally well developed. Many mammals do not have very acute vision. Most grazing animals, for example, have only fair eyesight, and some, such as rhinoceroses, are so nearsighted they can hardly see anything at a distance. Horses, cattle, and dogs seem to have little or no color vision. On the other hand, their hearing is excellent and their sense of smell frequently seems miraculous. We have only to think of a dog following the scent of an animal, picking it out of a maze of other, conflicting odors, or responding to sounds beyond the range of our hearing. The same holds true for animals such as deer, which stop browsing and lift their heads, poised for a fast getaway, when the air wafts along infinitesimal traces of scent and sound.

Birds, on the other hand, have practically no sense of smell at all. Among all the many thousands of bird species, only the kiwi relies on its "nose" and has very poor eyesight. All other birds have extremely keen eyesight, and some, such as hawks, possess what amounts to a kind of built-in telescopic lens that permits them to see fine detail hundreds of feet away. The fleet-footed birds—ostriches, rheas, and bustards—that live on the plains have very large eyes, and vie with the hawk in keenness of vision. Just as deer and antelopes depend mostly on their noses, so the ostrich relies mostly on its eyes.

In their natural habitat of the African plains, the common enemies of ostriches as well as of the large grazing mammals, such as zebras, gnu, and antelopes, are the powerful predators—mainly lions and leopards —that are native to those regions. None of these predators, except perhaps the cheetah, can overtake its prey if the animals are alerted and on the run. Only if the lion can creep up against the wind to a zebra, or under perfect cover to an ostrich, can it successfully make a kill.

Therefore, it is obvious that a pooling of "talents" by ostriches and grazing animals would be a highly effective means of increasing mutual protection against sneak attacks. From their imposing eight-foot height, ostriches scan the surrounding countryside with their

Zebras, which have a keen sense of smell, and ostriches, which have excellent vision, are often found together.

keen eyes, alert to the slightest suspicious movement. With their superior sense of smell and hearing, the zebras and antelopes catch any danger signal which the wind transmits. Together these animals have the eyes, ears, and noses that help them evade the enemy that continually lurks in the tall grass and dense undergrowth of the African bushveld.

Moreover, from the association with grazing animals, the ostriches receive a bonus which to them is most convenient. Very much in the manner of cattle egrets, ostriches watch for insects, reptiles, and rodents which are flushed from the grass by the movements of the grazing herds, becoming welcome prey to the big birds. Ostriches are omnivorous, eating any small creatures they can catch, in addition to the seeds, fruit, and leaves that generally make up the bulk of their diet.

In South America we find an analogous situation between the pampas deer and the rheas. Rheas are often called South American ostriches, which is a misleading term because they are not related to the African ostrich and are quite different in appearance and anatomical features. Rheas stand only about five feet tall, have three-toed feet—as compared to the two-toed feet of the ostrich—a feathered neck and thighs, and lack the distinctive tail plumes of the ostrich. However, they share with the African bird the ability to escape from predators by virtue of their keen vision and fleetness of foot. The association between rheas and pampas deer gives both partners the same advantages ostriches and grazing mammals derive from their relationship.

The greatest and most colorful variety of all types of symbiotic alliances is found among the lower animal groups. This holds true not only for the relationships that are mutually beneficial, but also for those essential to the survival of one or both partners. The ocean with its enormously diverse animal life has a wealth of relationships vital for survival, offsetting the deadly underwater battles for existence in which smaller creatures continuously fall prey to bigger ones.

In recent years underwater studies have been greatly facilitated by a number of devices, including those that permit man to stay below the surface for prolonged periods of time without being encumbered with heavy diving apparatus. The efficient breathing devices used today allow man to become temporarily one of the ocean's creatures, to mingle with them and observe their ways of life from close quarters in their own natural environment. Many facts that had never before been known have come to light in such studies, and among the most fascinating of all is the phenomenon called cleaning symbiosis by biologists.

Actually many symbiotic associations between land animals fall into the same category. The tickbirds that remove parasites from the skin of grazing animals; the Egyptian plover that picks leeches from the crocodile's hide and mouth; and the pseudoscorpion that eats the mites infesting the bodies of large insects are all examples of cleaning symbiosis. However, none of these can be said to be so essential to the "client" animal that its health would be gravely and perhaps critically affected without such services. Undoubtedly the animals experience relief and a greater measure of comfort at being rid of the parasites, but that generally seems to be the extent of the benefit they derive.

In the ocean, however, the situation is quite different. Not only has research disclosed that cleaning symbioses are much more widespread, involving apparently the majority of fishes, but the findings also indicate that cleaning is essential to the health of the fish being serviced, and that their numbers would be greatly decimated without it. Several biologists who have studied the phenomenon are convinced that the cleaning of fish by various marine creatures, including other fish and shrimp, has great ecological importance: it keeps huge numbers of fish healthy and thereby available as food for other animals, including man.

One of the first naturalists to study cleaning symbiosis among marine animals was the famous American zoologist William Beebe. About forty years ago, he

A labroid cleaner fish attends a butterfly fish.

observed small red crabs climbing on the bodies of marine iguanas, the large, ugly, but harmless, vegetarian lizards found on the Galápagos Islands. Taking advantage of the periods when the iguanas were sunbathing, the crabs foraged for ticks which embedded themselves in the leathery skin of the big reptiles.

Intrigued by the activities of the crabs, Beebe began to watch for other instances of symbiotic cleaning by marine animals, and a few years later was rewarded when he saw small wrasses, spiny-finned fish allied to the parrot fish, cleaning other fishes in the waters around Haiti. Beebe at that time wrote about his observations and compared the activities of the cleaner fish to those of the tickbirds and the Egyptian plover.

With the turbulent years of World War II intervening, it was not until the late forties and the following decade that intensive studies of underwater life were resumed. One of the first to investigate the significance of cleaning symbioses in the ocean was Dr. Conrad Limbaugh, a marine biologist with the Scripps Institute at La Jolla, California. Several years later a German scientist, Dr. Irenäus Eibl-Eibesfeldt of the Max Planck Institute in Bavaria, Germany, launched a long-term research project, conducted in suitable environments in both the Atlantic and Indo-Pacific waters. In his book *In the Realm of the 1000 Atolls*, Dr. Eibl-Eibesfeldt revealed that his studies indicated the existence of cleaning symbioses among fish on a scale that seemed almost incredible. Practically all free-swimming fish, he found, from relatively small species

A barberpole fish permits its mouth to be cleaned by a small labroid fish.

up to and including such monsters as sharks and rays, were involved.

In the meantime, other studies have supplemented Dr. Eibl-Eibesfeldt's findings. There are now approximately forty known species of cleaner fish, and more than half a dozen cleaner shrimp. Because they are so numerous, the cleaner fish are ecologically the more important of the two, although some of the shrimp have fascinating cleaning techniques.

One of the most interesting features of the cleaners, fish as well as shrimp, is their distinctive coloration. In most cases cleaners have very bright color patterns that contrast sharply with their surroundings. Regard-

A brightly colored labroid cleaner fish.

less of the background body color, most cleaner fish are barred or striped. This seems to be so universal a pattern among these fish that biologists speak of a "uniform" worn by cleaners which clearly advertises them as such. The background colors vary according to the regions in which the fish occur. In the Indo-Pacific waters the color tends to be blue, while many cleaners in the Caribbean are yellow. Both groups, however, have dark lateral bands running the entire length of the body from head to tail fin. This band, then, seems to be the identifying insignia of the cleaners.

There can be no doubt that recognition of cleaner by client fish is by visual means. Fish have good vision, apparently within the same color range as human beings—from red through violet—a fact that does not apply to a great number of other animals. The conspicuously colored and patterned cleaners can be sure that their clients will recognize them so that they do not run the risk of winding up in the latters' stomachs —victims of mistaken identity.

That the fish which depend upon the services of cleaners do indeed recognize them visually is further demonstrated by the presence of mimics that look strikingly like the cleaners but are in fact predators waiting for a chance to attack. One such predator, called

Aspindontus taeniatus, is a mimic of labroid cleaner fish found in the Indian Ocean. Wearing the same blue, black-barred "uniform," this predator approaches a larger fish, imitating the cleaner's distinctive movements. Once the impostor gets close enough to strike, however, all pretense is dropped, and the predator, which has an underslung jaw and sharp long teeth, lunges at the unsuspecting victim and tears large chunks of flesh from its body.

The manner in which cleaner fish approach their clients varies considerably. In the more northern waters, the ceremonies that precede the cleaning process tend to be quite simple, even casual. The cleaner— a small wrasse, for example, or a diminutive goby— swims up to a larger fish, and upon getting a responsive signal, begins to hunt for various small parasites, fungus growths, sores, and damaged tissue on the body of the customer.

Two genuine cleaners (center and bottom) and a predatory mimic (top).

In the tropical oceans, however, the procedure is usually quite different. The brilliantly colored cleaner fish of those regions may put on displays that take as much time and are as elaborate and intricate as many courtship rites. Performing a regular ballet around their prospective client, the cleaners will dart forward, flip sideways, and dance up and down while extending their fins and beating their tails. This performance is repeated until the customer gets the message and slows down.

Now the client fish responds to the signals by approaching the cleaner, pausing, and then often assuming peculiar positions that indicate its willingness to be cleaned. Some roll over on their sides, others will stand on their heads or float at awkward angles. If, however, the customers outnumber the cleaners, the ceremonies may be marred by fights that erupt among the client fish, each of which wants to be first in line.

Finally the cleaning process can begin. Starting at the head, and then working its way along the sides of the body toward the tail, the cleaner nibbles with its small sharp teeth at the various fungus growths, ulcerated sores, and parasites—mainly minute crustaceans—which commonly trouble fish. Some, such as certain species of discus fish, are plagued more than others by fungus growths.

Cleaners are careful and thorough. A cleaner never seems to hurt its customer, but indicates with gentle nudges when it is time to raise a fin so that the areas below can be properly cleaned. It also will not hesitate to enter the mouth and the gill openings of the larger fish, and frequently works its way along the jaws, between the teeth, and even deep down into the throat. Probably the most amazing fact about the phenomenon of cleaning symbiosis is that, according to all evidence, even such fearsome predators as shark and barracuda never eat the cleaner fish, or even attempt to harm them.

The customers, which appear to be in a coma-like trance during the cleaning process, generally wait patiently until the cleaner is finished with its task. Some fish change color while they are being cleaned. No reason for this has yet been found, but it is a fact that a number of different species undergo drastic color changes. An almost black reef fish in the Indian Ocean turns light blue; several other species "blush," one turning pink, another, mottled red. One species of discus fish turns black, thus making the light-colored fungus growths stand out as white blotches on its scales.

When a client fish whose cleaner has been working inside its mouth feels the job is done, it will gently close and open the mouth to indicate to the cleaner that it should leave. Even when danger threatens, however, the client fish will remember to spit out the cleaner unharmed before it takes flight.

Basically the same rules that apply to cleaner fish also apply to shrimp doing that type of work. However, it appears that, in some cases at least, these relationships are still imperfectly established, for it has been observed that occasionally a fish eats the shrimp, even though such occurrences seem to be more or less accidental.

Cleaner shrimp, like their fish colleagues, are brightly colored and patterned, but do not have any special insignia similar to the lateral bands of so many cleaner fish. The elegant Pederson's shrimp, for instance, has a pattern of white stripes and purple spots on a transparent body, and the red-backed cleaner shrimp is ochre yellow with a brilliant red stripe, divided by a white dorsal line, running along the back. The boxer shrimp, on the other hand, is white but banded with bright red. Some of the cleaner shrimp have long white antennae which they wave about like flags when they wish to attract a customer.

The cleaning of large fish always contains a dramatic element for the observer. Only a very blasé person could fail to be fascinated on seeing a tiny shrimp move boldly into the gaping mouth of a large moray eel—one of the most rapacious of all marine predators—and work its way around the sharp, saber-like teeth and into the throat of this monster.

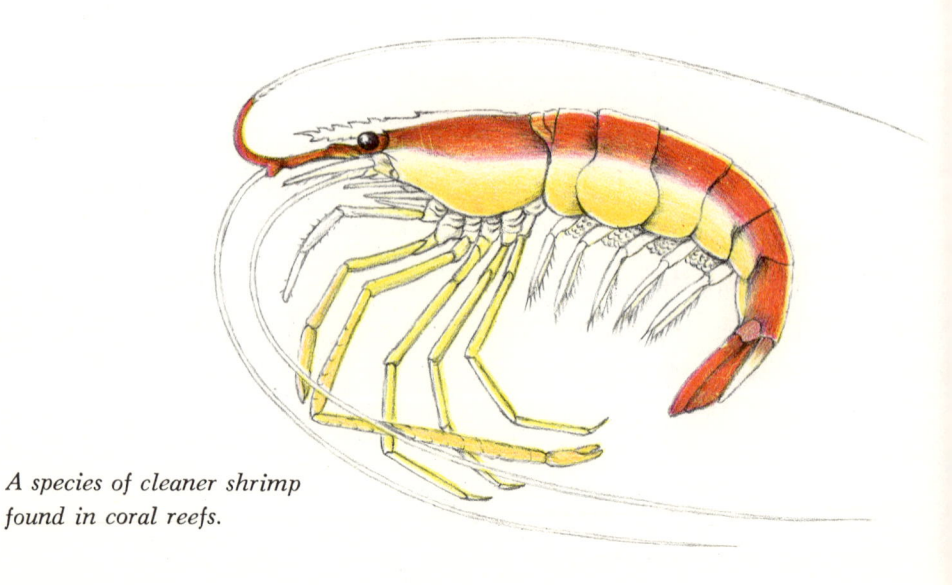

A species of cleaner shrimp found in coral reefs.

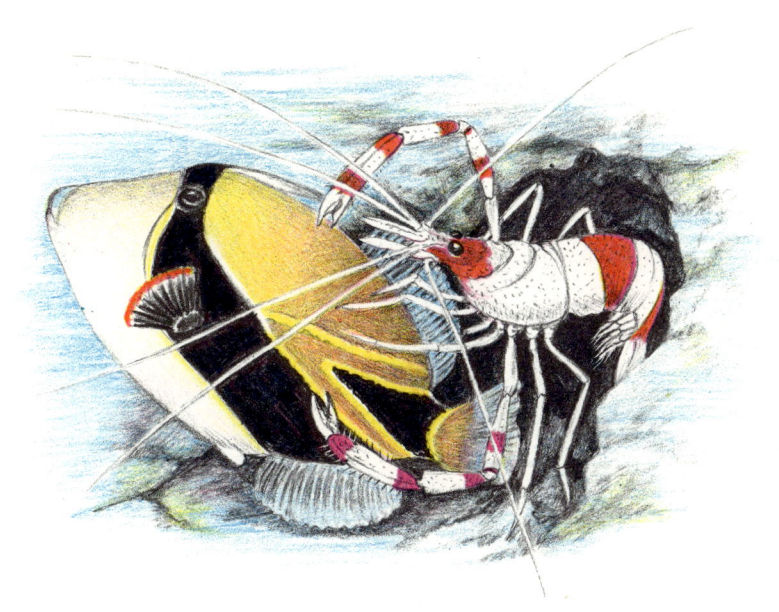

A client fish backs into the underwater home of a cleaner shrimp.

Some fish search out the cleaner shrimp in the latters' homes in cracks and hollows among rocks. In order not to scare the cleaners and to reassure them about the peaceful intent of the visit, the polite clients will back in tail first, thereby signaling that they wish to be attended.

During the extensive studies by Limbaugh and Eibl-Eibesfeldt, it was discovered that fixed locations in many parts of the oceans are established as "cleaning stations." Fish that wish to be serviced show up at these stations, often in great numbers, and are attended by what we may call the "resident cleaners." Dr. Eibl-Eibesfeldt noted one entire school of fish, which all signaled their cleaners by opening their mouths, turning their lips outward, and spreading their gill covers.

After they had been cleaned to their satisfaction, they closed their mouths and took off with a flip of their tails.

At the cleaning stations, fish are processed at a truly incredible rate. Dr. Limbaugh found that a single cleaner attended three hundred clients during a six-hour period. Very sick or injured fish would return to the station several times during the same day to receive repeated "treatments."

Interested in finding out exactly how important the cleaning is for the fish, and whether the lack of this service has any noticeable results, Dr. Limbaugh removed all the cleaners he could find from a few small coral reefs in the Bahamas. The results were dramatic and drastic. Within just a few days, the fish population was greatly reduced, and two weeks later, all except those that habitually live in the reefs were gone. Among the latter, many had developed sores and swellings as well as tattered fins, telltale signs of ill health. Dr. Limbaugh concluded from these observations that cleaning is indeed vital to the health of fish. It is believed that many of the famous fishing grounds, the places where fish are traditionally found in huge numbers, are actually cleaning stations to which they flock in order to be serviced.

Although the importance of symbiotic cleaning in the ocean cannot yet be fully appraised, it appears certain that this phenomenon has had considerable effects on the evolution of marine organisms, thereby influencing the ecology of the entire ocean.

Symbiotic cleaning is probably the most widespread type of cooperation for survival found in the ocean, but it is by no means the only one. As mentioned in the preceding chapter, many relationships, especially those between sea anemones and other marine creatures, have evolved to the point where a dependency has been established. Such is the case with the anemone shrimp, which lives among the tentacles of the annulated anemone. If the shrimp is removed, the anemone dies shortly afterward. The shrimp evidently is needed to supply the anemone with food, although other factors may also play a part. The shrimp, of course, gets considerable protection by living among its partner's stinging tentacles.

A similar alliance exists between another species of anemone and its symbiotic fish. Experiments in the aquarium have shown that the anemone soon dies after the fish is removed. Here, too, further research may disclose additional facts about the reasons for the anemone's dependency.

Probably the most intimate and completely dependent association between a sea anemone and another animal is that between *Adamsia palliata* and the hermit crab *Eupagurus prideauxi*. This anemone can live only with the crab, and dies if it is removed from the latter's shell. The crab also needs its partner for a successful life. All attempts at keeping the two alive and well without each other have failed so far.

Returning from the life of the ocean to the world of insect life, we find a great many mutually advantageous relationships essential to the survival of one or both partners. Here again, the social insects are primarily involved in these associations.

An outstanding example of a totally dependent ant "guest" is a small beetle that has become completely adapted to life in the ant community. Blind and wingless, it can survive only if the ants feed it. This they do faithfully, because the beetle gives something in return. It has special glands that secrete, through short tufts of hair called trichomes, a substance which the ants relish greatly. The association is thus a luxury for the ants, but a necessity for the beetle.

Some of the most unusual mutualistic alliances that have become essential to one of the partners are found to exist between ants and certain species of small blue butterflies. As mentioned earlier, the majority of the blues have some form of symbiotic relationship with various ant species. In most cases it is just a matter of

Caterpillar of the large blue butterfly on wild thyme.

The caterpillar is transported underground by an ant.

the ants' visiting the butterflies on their food plants and "milking" them of the sweet honeydew they secrete, while at the same time protecting them from enemies. A few blues, however, have gone far beyond that stage and cannot complete their metamorphosis without the help of the ants.

An example of such a dependency is found in the life cycle of the European butterfly called the large blue, with a wingspread of about one and one-half inches, larger than the majority of its relatives.

In the spring the female deposits her eggs on wild thyme. After hatching, the caterpillar, like all butterfly larvae, begins to feed on the leaves of its food plant. This lasts until after the second molt, the skin shedding which permits insect larvae to grow. By that time the caterpillar, which looks like a grub, has developed a special honeydew gland on its back. Ants quickly discover the gland and start visiting and "milking their cow" on the thyme. After the fourth molt, the larva stops eating the leaves, and instead wanders about aimlessly. Sooner or later, ants find the caterpillar and carry it to their underground nest. The caterpillar actually prepares itself for this transport, hunching its back to make it easier for the ant to carry the load.

In the dark underground chambers of the ant colony, a strange new life begins for the caterpillar. It seems as though the change of locality has caused a change of personality too, for it suddenly turns into a carnivorous animal. Instead of the thyme leaves that were formerly its exclusive food, it now dines on tiny ant larvae.

While the ants could normally be expected to become enraged over such attacks on their brood, in this case they ignore the assault. Instead they continue to

Despite the fact that the caterpillar feeds on small ant grubs in the ants' underground nest, it is a welcome guest because of its sweet secretions.

crowd around and stimulate the caterpillar to give up more of its honeydew.

By the time the cold season arrives the caterpillar has grown to roughly four times its original size. It now stops eating and begins to hibernate, which means spending the winter in a lethargic state in which most body functions slow down or halt altogether.

In the spring the onset of warmer weather brings the caterpillar out of its hibernation. It resumes preying on the ant brood, and again yields honeydew whenever stroked by the adult ants. After a while, it has attained full growth and is ready to change into a pupa, which is attached by its tail end to the roof of the ant chamber in which the caterpillar has spent its time of growth. The pupa soon falls to the ground and remains there, untouched by the ants, until the butterfly emerges, makes its way through the ants' passages to the world of sun and blue skies, and starts new generations of blues on their curious life cycle.

Studies have shown that this butterfly could not complete its metamorphosis without the help of the ants. Not only does it need ant larvae as food at a certain stage of its development, but it also seems to need the stimulation of the honeydew gland for proper growth.

Probably the most famous of all symbiotic relationships involving social insects are those between nectar-seeking insects and flowering plants. As mentioned briefly at the beginning of this chapter, these relationships are of prime ecological importance and have a vital role in our agriculture. Without the good services of flower-seeking insects, and especially honeybees, a great part of the harvest of fruit we reap today would be lost. Beehives are regularly rented out to orchard owners in the spring in order to assure proper pollination of apple, peach, pear, and other fruit trees.

In many tropical regions, fertilization of flowers is achieved mainly by small birds, such as hummingbirds, sunbirds, and honey creepers. These birds all have long slender bills with which they can reach into the long tubular blossoms that are typical of many tropical plants. The bills of hummingbirds especially show a wide range of adaptations. One species, appropriately called the sword bearer, has a bill that is longer than its body. The bills of other species are sickle-shaped, and all have tongues that can be extended far beyond the tip of the bill. As they insert the bill in the flower to withdraw the nectar, they inadvertently fertilize it.

A similar function is performed by certain moths. A group called sphinx or hawk moths, which have very long "tongues," or sucking tubes, are of major importance in the pollination of a great number of flower-

Honeybee collecting nectar and pollen in a sage blossom.

ing plants. In some cases, highly specialized relationships have developed, with a plant becoming entirely dependent upon fertilization by one particular species of moth. One of the most intriguing nature "detective" stories concerns one of these moths. In Madagascar naturalists puzzled over an orchid that had its nectaries hidden at the bottom of the blossom more than ten inches long. They could not find any animal, whether bird or insect, that was capable of pollinating this orchid. The English naturalist Alfred Russell Wallace predicted in 1891 that a hawk moth with a tongue long enough to reach inside these blossoms must exist, because only such a moth could have a tongue of that kind.

Eleven years later, the moth was discovered. Not surprisingly, the word *predicta*—the predicted one—was tacked on to the insect's scientific name.

Although many sphinx moths fly by day, when they are often mistaken for hummingbirds because of their large size and hovering flight, most are active only at dusk. In perfect coordination, the plants they visit, such as some species of honeysuckle, give off their strongest scent in the evening, thereby attracting their guests.

Such highly specialized symbiotic alliances between just one kind of plant and a single species or genus of insect have evolved in a number of cases, and are fascinating in the complexity of their operation. Probably one of the most interesting is the mutual dependency between the yucca plant and the small yucca moths.

Yuccas, also called Spanish dagger plants, are a common sight on the arid plains of the American West and Southwest. The white flower is the state flower of New Mexico. The plant has long, stiff, dagger-like leaves at the base. From this base rises a long stem bearing at the top a large cluster of white flowers.

The yucca moths, which comprise a small and exclusively American family, are also white, although a few have some dark markings. The best-known species, *Tegeticula yucasella*, can serve as an example for all of them, because their habits are similar.

All activities take place at night. During the daytime, the moths rest in the half-closed flowers, thus escaping the worst heat. At dusk, they begin to get active. Flitting from flower to flower, they collect loads of pollen. This work is done only by the females, which have special anatomical adaptations that enable them to perform efficiently.

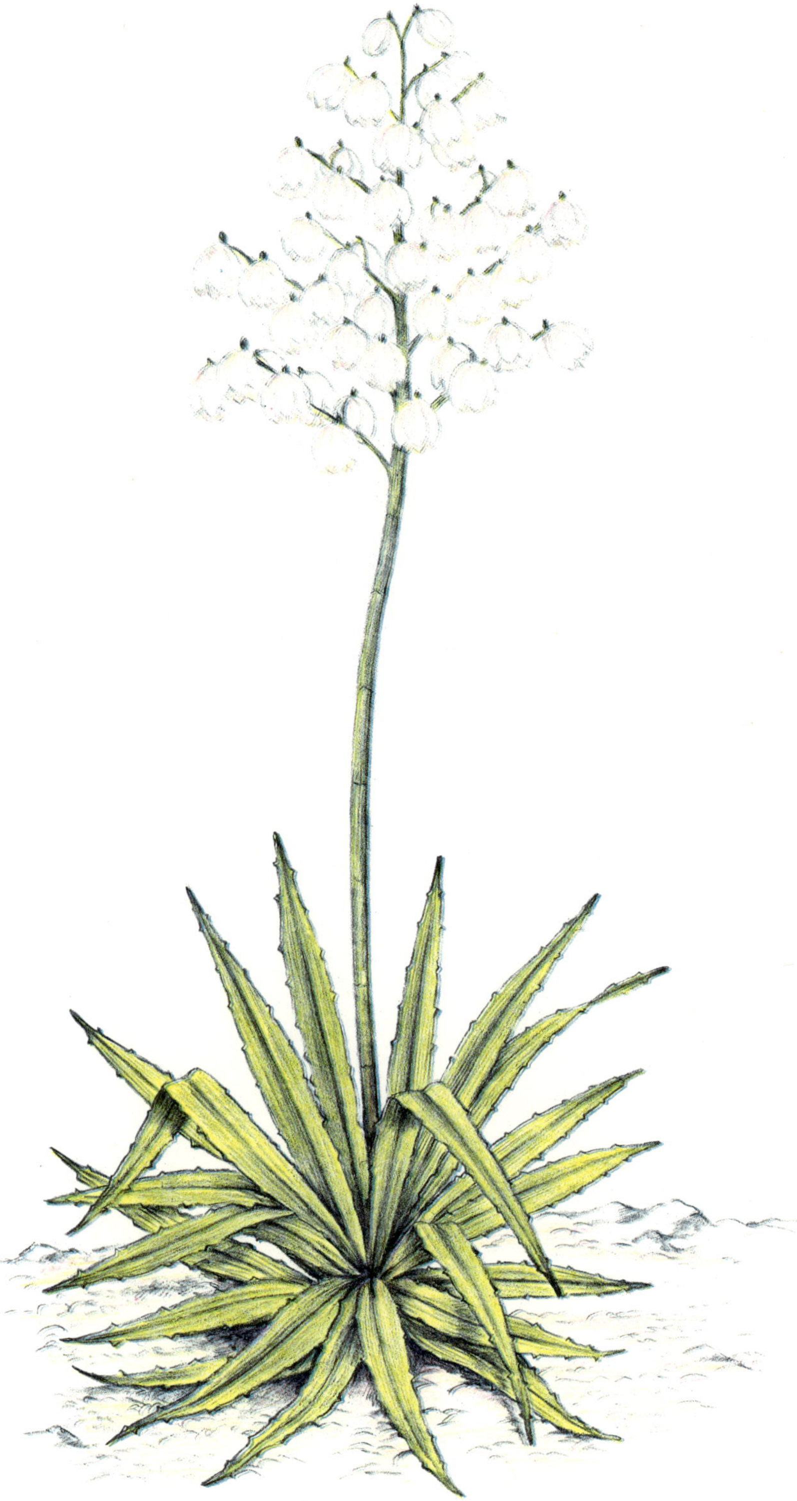

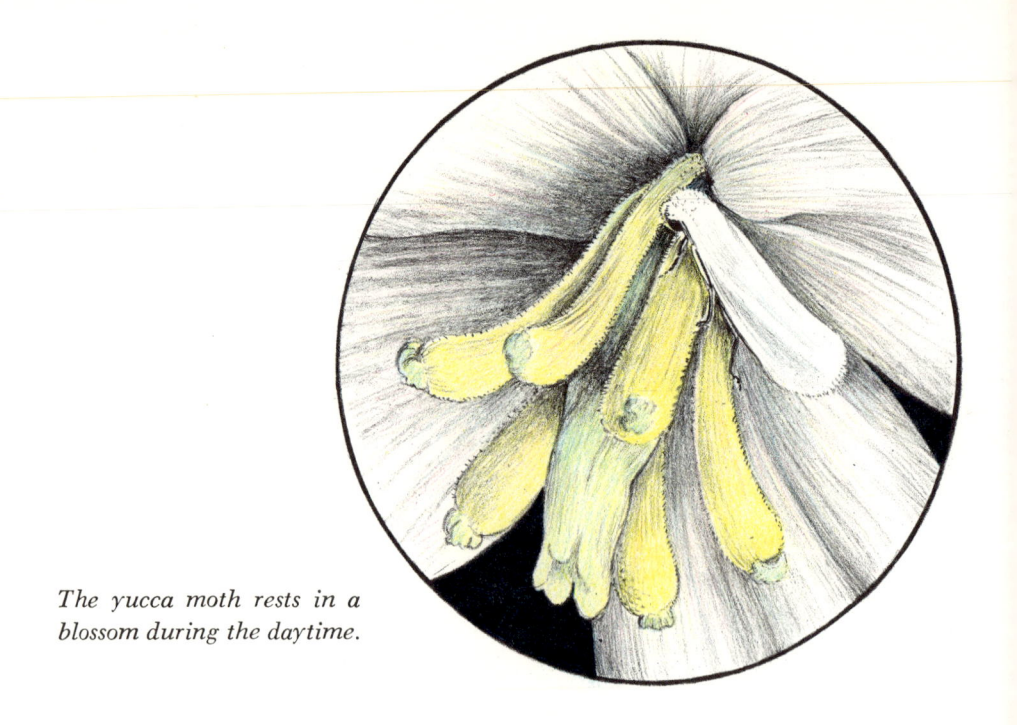

The yucca moth rests in a blossom during the daytime.

After locating a suitably fresh flower, the moth takes hold of the stamen, runs up to the top to the anther and moves her head back and forth, removing the pollen with her mouth parts and shaping it into a pellet with her forelegs. After collecting pollen from several flowers until she has formed a load about three times as big as her head, she positions this ball firmly against her neck with the help of the front trochanters (a trochanter is the second division, counting outward from the body, of an insect's leg).

With the load of pollen firmly wedged against her body, the moth flies to another flower, deposits her egg in the latter's ovary, climbs up the style, and packs her load of pollen into a depression on the upper part of the stigma.

Having thus fertilized the flower, the moth has enabled the yucca's ovary to become a fruit with many flat seeds. When the moth larvae hatch, they are tucked snugly inside a "nursery" stocked with seeds upon which they can feed. However, because the moth deposits only one or two eggs in each flower, only a few seeds are eaten: this is the price paid by the yucca for the invaluable services of the little moth. After the caterpillars have completed their growth, there are still more than enough seeds left for new yucca plants to develop.

The action of the moth, both in gathering and depositing the pollen can only be described as "deliberate," even though the behavior does not involve conscious reasoning as in human deliberation. The moth's acts are not merely coincidental by-products of the insect's usual daily activities, such as the fertilization of flowers by insects searching for nectar. The instinct that tells the yucca moth exactly what to do works to perfection, to the point of making sure that older flowers not "open for business" are avoided.

Hardly less amazing than the case of the yucca moth is that of the fig gall-wasp. This tiny creature is the only agent capable of pollinating the common fig tree.

In order to understand the full complexity of this relationship, and the marvelous balance maintained to satisfy the needs of both the tree and the wasp, it is best to follow the entire process on a wild fig tree. These trees still exist in some regions of Italy. They are monoecious, meaning that both male and female flow-

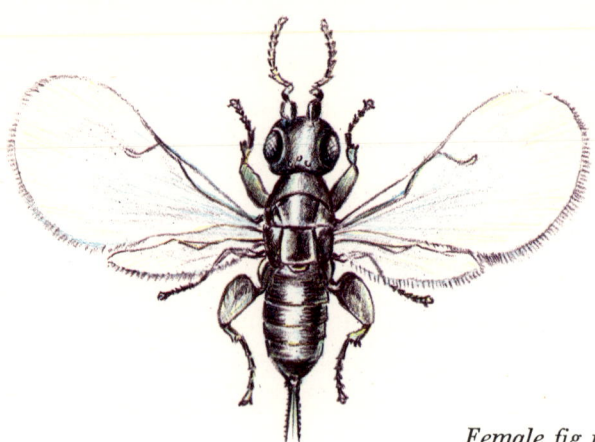

Female fig wasp, greatly enlarged.

ers are found on the same plant. In the spring the trees produce small, inedible figs called profichi. They have both male and modified female flowers with very short styles. The female fig wasp deposits her eggs in these flowers, which thereupon form galls, round swellings of plant tissues. Inside the galls, the larvae develop.

In June and July the adult wasps emerge from the pupal state. The wingless males crawl into galls that contain females and mate with them. The females then make their way out of the gall and into the open, getting powdered with pollen as they pass the male flowers.

In the meantime the tree has formed large, edible figs that consist of female flowers only and are called fichi. The wasps try to deposit their eggs in these flowers. However, they have long styles that are closely bunched together, and resist every effort by a female wasp to place her eggs inside. Frustrated in her attempts, she nevertheless inadvertently fertilizes the flowers with the pollen she picked up on her way out of the profichi.

So far, so good. But if provision were not made for the new generation of fig wasps, there would not be any adults about to fertilize the following summer's crop of fichi. This important matter is taken care of in the fall, when a third type of figs is produced by the trees. They are appropriately called mammes, or mother figs, and are excellent incubators for wasp larvae. After depositing their eggs in the mamme flowers, the old female wasps die. The larvae hatch, hibernate in their snug nurseries, and finally emerge in the spring, ready to begin the cycle all over again.

Cultivated fig trees do not have male and female flowers on the same tree. Instead some trees produce only the edible, female flowers, commonly called Smyrna figs, and others produce only the profichi, which are known as caprifigs. For centuries before the nature of the symbiotic relationship between the fig plants and the wasp was understood, fig growers had planted a few caprifig trees among the Smyrna variety, or even hung a few branches of these trees, which for all purposes are male, among the others. This process is called caprification. We must assume that the separation of the monoecious wild fig trees into the male and female trees occurred thousands of years ago in Asia Minor, where the ancient Greeks grew fig trees from cuttings and finally succeeded in developing separate male and female trees.

When Smyrna fig trees were imported into California in 1880, fruit growers still did not know that male trees had to be planted among the others in order to get fruit. So the trees remained sterile until, years later, the caprifigs with their resident wasps were introduced. After that the trees yielded rich harvests year after year.

In recent times a seedless fig that does not need any fertilization has been developed. Needless to say it can be grown only from cuttings. The majority of high-quality figs that can be easily preserved still can be obtained only with the help of the little wasps.

In all the cases of symbiosis between plants and insects described thus far in this chapter, the plants' reproduction is assured only by the collaboration of their insect "partners." However, there are some interesting examples in which the very existence of the individual plant depends upon the good services of one particular kind of insect. In order to get the protection it needs,

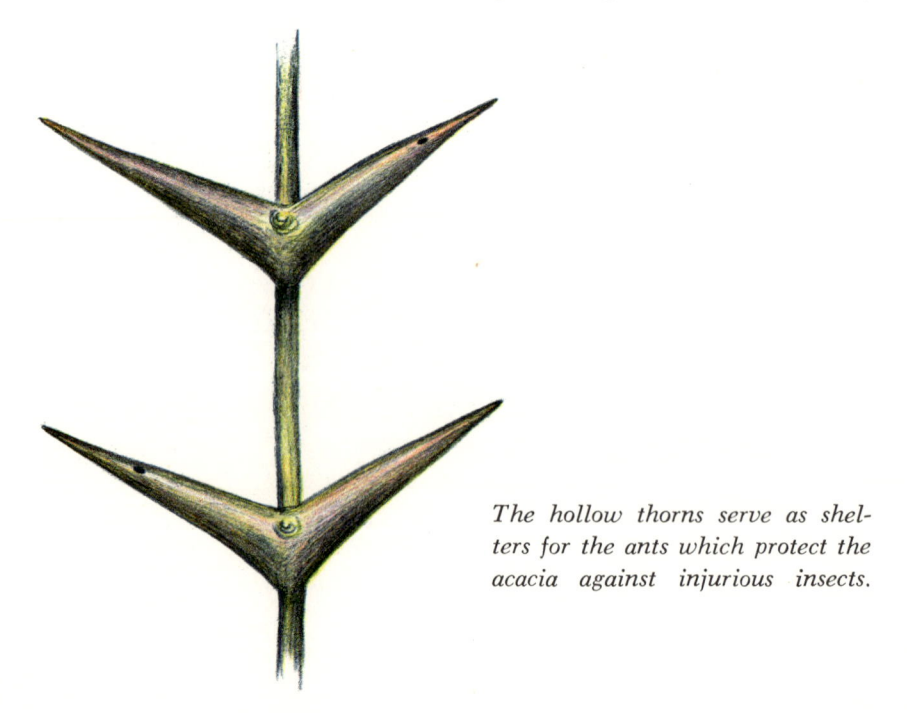

The hollow thorns serve as shelters for the ants which protect the acacia against injurious insects.

The small, fruitlike bodies at the tips of these acacia leaves are eaten by the ants that live in the acacia thorns.

the plant offers its partner a rich compensation: delicious food and adequate shelter.

Uusually ants are the insects involved in this type of symbiosis. There are a number of mostly tropical plants that are known for the accommodation they provide for ants. In one case the ants live in the hollow stems of the plant and feed on small clusters containing nutritious substances that develop on the segments of the stems. In exchange, the ants serve as a kind of police that keep away injurious insects.

Most famous for relationships of this type are several tropical species of acacia. *Acacia spherocephala,* for example, has long hollow thorns in which a species of ants builds its colonies. On the tree's smaller branches are nectar glands which secrete a sweet liquid eagerly

Ants on Acacia collinsii, *which provides its ant guests with food in the form of small nectaries at the base of the leaves.*

sought by the ants. Moreover, small appendages on the tips of the individual leaves provide additional food for the tree's insect tenants. In return, the ants patrol the branches of the acacia, chasing away or killing herbivorous insects which otherwise would defoliate the acacia.

Acacia collinsii offers its ant population a slightly different arrangement. Here again, large hollow thorns provide living space for the ants with which the tree lives in symbiotic alliance. However, the nectaries are arranged in rows of four near the base of the leaf stem, just where they join the branch. The "resident" ants of the species *Pseudomyrmex ferruginea* are very efficient plant pest controls, and successfully keep away insects that endanger the acacias. It seems that an unprotected tree is beset by insects which eat the shoots and damage it so severely that it cannot survive. In experiments in which biologists removed the ants and prevented them from returning to the trees, the acacias died within three to twelve months.

Partners and guests, parasites, passengers, and hangers-on—associations for convenience, gain, and survival—we have encountered every type of symbiotic relationship in this book. We find them among a host of different types of creatures, and in every conceivable environment: in the forests and on the plains, high up in trees and down in the shadows of the jungle floor, underground and underwater. Diverse as they are, they have one thing in common. As nothing else can, they emphasize, in the words of Dr. Conrad Limbaugh, "the role of cooperation in nature as opposed to the tooth-and-nail concept." This cooperation is a vital part of the balance of nature, whose lessons man can ignore only at his own hazard.

Index

(*Page numbers in italics refer to illustrations.*)

Acacia: hollow thorns of, *118*, 119, 120; leaves of, *119*
Acacia collinsii, 120, *120*
Acacia spherocephala, 119
Actiniaria, 70
Adamsia palliata, 105
Africa: bushveld in, 93; carmine bee eater of, *26*, 27, 28, 29; cattle egret of, 25; crocodile of, 67, *68*, 69, 70, 94; honey guide of, 61, *63*, 63–66; Kori bustard of, *26*, 27–28, 29; mongoose of, 67, 68; pompadoured hornbill of, 31; ratel of, 61–63, *62;* trochilus of, 67, *67*, *68*, 69–70, 94
Antelopes, 92, 93
Ants, *11*, 17, 32, 39–40, 54, 83–87, 106, 118–20; and acacia, 119–20, *120;* and aphids, 14, 84, *84;* and crickets, 54, 55; and larvae of blue butterflies, 86–87, *107*, 107–109, *108;* and larvae of flies, 56; and *Lomechusa*, 85; and mites, 55, 55, 56; and silverfish, 55, *55*
Aphids, and ants, 14, 84, *84*
Arachnoidea, 79
Asia, cattle egret of, 25
Aspindontus taeniatus, 99, *99*
Australian parakeet, 38

Bacteria, 18, 19, 20
Balance of nature, 12
Barber ants, *11*, 83
Barberpole fish, 97
Barracuda, 101
Beebe, William, 95, 96
Bee eater, carmine, *26*, 27, 28, 29
Bee louse, 54
Bees: honey, 17, 53, 54, 62, 90, 110; solitary, parasitized by oil-beetle larvae, 52, 53
Birds: flowers fertilized by, 110; insect-eating, 27, 29; keen eyesight of, 29, 91
Blue butterfly, 86, *86*, 106, 107; larvae of, and ants, 86–87, *107*, 107–109, *108*

Bower bird, 60
Boxer shrimp, 102
Bubonic plague, 16
Buffalo, 12, 24
Bumblebee, 51–52, *52*
Bustard, Kori, *26*, 27–28, *29*
Butterfly, blue. See Blue butterfly.
Butterfly fish, *95*

Carmine bee eater, *26*, 27, 28, *29*
Caterpillars: and ants, 86–87, *107*, 107–109, *108;* beehives invaded by, 54
Cattle, 19, 24, 27, 58, 91
Cattle egret, *22*, *23*, 24–26
Cellulose, bacterial action upon, 19
Cheetah, 92
Cleaner fish, *95*, 96, 97, *97*, 98, 99–101
Cleaning symbiosis, 83, 94; in ocean, 95–104
Clownfish, 72, 74, *97*
Commensalism (eating together), 23, 24–37
Cooperation in nature, emphasized by symbiosis, 121
Cow bird, 27
Crabs: hermit, *73*, *74*, 74–77, *75*, 105; and marine iguanas, 96
Crickets, 50; and ants, *54*, 55
Crocodile, 67, *68*, 69, 70, 94
Crocodile bird. See Egyptian plover.
Cuckoo, 64

Darwin, Charles, 11, 14
Deer, 91; pampas, and rhea, 93
Discus fish, 100, 101

Driver ants, 32
Dung beetles, 47–51, *48*

Echinoderms, 41
Ecology, 12, 20
Ectoparasites, *15*, 16
Eel, 41
Egret, cattle, *22*, *23*, 24–26
Egyptian plover, 67, *67*, *68*, 69–70, 94
Eibl-Eibesfeldt, Irenäus, 96, 97, 103
Elephant, 12, 24
Endoparasites, 16, *16*
Eupagurus prideauxi, 105
Evolution, 11, 12, 14

Fierasfer acus, 41, 42, *42*, 43, *43*
Fig gall-wasp, 115–17, *116*
Fish, good vision of, 98
Fish hawk, 38
Fleas, *15*, 16
Fluke, as endoparasite, *16*
Food chain, 11, 12

Galápagos Islands, 96
Gnu, 92
Gorilla, 12
Grackle, 38
Grasshopper, 28
Grazing mammals, associated with ostriches, 92–93
Guenon monkey, 31
Gull, 27

Harlequin beetle, 79, *80*, 81, 82
Hawk, 29, 91
Hawk moth, 110, 111
Hermit crab, *73*, *74*, 74–77, *75*, 105

Herodotus, 67
Holothurians, 41
Honey badger. See Ratel.
Honeybees, 17, 53, 54, 62, 90, 110
Honey creeper, 110
Honey guide, 61, *63*, 63–66
Hookworm, 16, *16*
Hornbill, pompadoured, *30*, 31
Hummingbird, 110, *111*
Hypnotoxin, 71

Iguana, marine, 96
In the Realm of the 1000 Atolls (Eibl-Eibesfeldt), 96
Indian rufous woodpecker, 38–40, *39*
Indicatoridae, 63
Insects: parasitoid, 17, *17*; sensory organs of, 50; social, 17, 54, 106
Intestinal microorganisms, 18, *19*, 20

Jellyfish, 77, *78*
Jungle Book (Kipling), 69

Kipling, Rudyard, 69
Kiwi, 91
Kori bustard, *26*, 27–28, 29

Labroid cleaner fish, *95*, *98*, 99
Larvae: blue butterfly, and ants, 86–87, *107*, 107–109, *108*; defined, 41; fly, and ants, 56; mite, and dung beetles, 49–50, 51; oil-beetle, and solitary bees, 52, 53, *53*
Leopard, 92

Leptothorax provancheri, *11*, 83
Lice, 16
Limbaugh, Conrad, 96, 103, 104, 121
Lion, 92
Liver fluke, *16*
Lomechusa, 85
Longhorn beetle, 79

Mammals, as advanced group, 90–91
Man-of-war fish, 77, *78*, 79
Microorganisms, 18, 19; intestinal, 18, *19*, 20
Mites, 79, 81, 82, *82*; and ants, 55, *55*, *56*; larvae of, and dung beetles, 49–50, 51
Mongoose, 67, 68
Monkey, guenon, 31
Moray eel, 102
Moths, 31, 110, 111, 112, 114–15
Mutualism, 57, 59, 71, 87, 89, 106
Myrmecophiles, 55
Myrmica brevinodis, 83

Nematocyst, 71, *71*, 74
Nomeus gronovii gmelin, 77, *78*, 79
North America, cattle egret of, 25

Octopus, 76
Oil-beetle larvae, 52, 53, *53*
Osprey, 38
Ostrich, 91, *92*, 92–93
Otter, 60
Oxpeckers. See Tickbird.
Oysters, and pearlfish, 44

Pagurus arrosor, 73, 74, 74–77, 75
Parakeet, Australian, 38
Parasitism, 15–16, 18; social, 17, 23, 54–57, 59
Parasitoid insects, 17, *17*
Pearlfish, 44
Pederson's shrimp, 102
Phoresy (transporting passengers), 23, 24, 47–53
Physalia arethusa, 77, *78*
Pilot fish, 33, *34*, 35, *35*
Pliny the Elder, quoted, 67
Plover, Egyptian, 67, *67*, *68*, 69–70, 94
Pollination, 110, 114, 115
Pompadoured hornbill, *30*, 31
Pontonia, 45, *45*
Portuguese man-of-war, 77, *78*, 79
Predatory mimic cleaner fish, 99, *99*
Pseudomyrmex ferruginea, 120
Pseudoscorpion, *81*, 81–82, *82*

Radianthus ritteri, 72, 74
Ratel, 61–63, *62*, 65
Ray (fish), 97
Red-backed cleaner shrimp, 102
Remora, *34*, 35–36, *36*, 37
Rhea, 91, 93
Rhinoceros, *22*, 24, 91
Rhinoceros beetle, *49*

Scarab, 47–49, *48*
Sea anemone, 70–77, *73*, *74*, *75*, 105
Sea cucumber, 41–42, *42*, 43, *43*, 44

Sea urchin, 45, 46, *46*
Sharks, 33, *34*, 35, 36, 37, 97, 101
Shearwater, sooty, 37
Sheep tick, *15*
Shrimp, 44–45, *45;* anemone, 105; cleaner, 97, 101–103, *102*
Silverfish, 55, *55*
Social insects, 17, 54, 106
Social parasitism, 17, 23, 54–57, 59
Sooty shearwater, 37
South America, rheas and deer of, 93
Sphinx moth, 110, 111
Sprengel, Christian Konrad, 14
Sunbird, 110
Symbiosis, 13, 14, 18, 19, 20, 23; types of, 23–24. *See also* Cleaning symbiosis; Mutualism.
Synoecy (sharing shelter), 23, 24, 37–44

Tapeworm, 16
Tegeticula yucasella, 112
Termites, 17, 19, *19*, 38, 85
Tick, *15*
Tickbird, *58*, *59*, 94
Trichomes, 106
Trochanters, 114
Trochilus (Egyptian plover), 67, *67*, *68*, 69–70, 94
Tuatara, 37
Tumblebug (scarab), 47–49, *48*

Urchin fish, 45, 46, *46*

Wallace, Alfred Russell, 111

Wasps: fig, 115–17, *116*; parasitoid, 17, *17*
Woodpecker, Indian rufous, 38–40, *39*
Worms, and dung beetles, 49, 50–51

Wrasse, 96

Yucca moth, 112, 114–15
Yucca plant, 112, *113*, 114, 115

Zebra, 29, *92*, 93